Lecture Notes in Mathematic

T0281318

Editors:
J.-M. Morel, Cachan
F. Takens, Groningen
B. Teissier, Paris

FONDAZIONE CIME
ROBERTO CONTI
CENTRO INTERNAZIONALE MATEMATICO ESTIVO
INTERNATIONAL MATHEMATICAL SUMMER CENTER

C.I.M.E. means Centro Internazionale Matematico Estivo, that is, International Mathematical Summer Center. Conceived in the early fifties, it was born in 1954 and made welcome by the world mathematical community where it remains in good health and spirit. Many mathematicians from all over the world have been involved in a way or another in C.I.M.E.'s activities during the past years.

So they already know what the C.I.M.E. is all about. For the benefit of future potential users and co-operators the main purposes and the functioning of the Centre may be summarized as follows: every year, during the summer, Sessions (three or four as a rule) on different themes from pure and applied mathematics are offered by application to mathematicians from all countries. Each session is generally based on three or four main courses (24−30 hours over a period of 6−8 working days) held from specialists of international renown, plus a certain number of seminars.

A C.I.M.E. Session, therefore, is neither a Symposium, nor just a School, but maybe a blend of both. The aim is that of bringing to the attention of younger researchers the origins, later developments, and perspectives of some branch of live mathematics.

The topics of the courses are generally of international resonance and the participation of the courses cover the expertise of different countries and continents. Such combination, gave an excellent opportunity to young participants to be acquainted with the most advance research in the topics of the courses and the possibility of an interchange with the world famous specialists. The full immersion atmosphere of the courses and the daily exchange among participants are a first building brick in the edifice of international collaboration in mathematical research.

C.I.M.E. Director
Pietro ZECCA
Dipartimento di Energetica "S. Stecco"
Università di Firenze
Via S. Marta, 3
50139 Florence
Italy
e-mail: zecca@unifi.it

C.I.M.E. Secretary
Elvira MASCOLO
Dipartimento di Matematica
Università di Firenze
viale G.B. Morgagni 67/A
50134 Florence
Italy
e-mail: mascolo@math.unifi.it

For more information see CIME's homepage: http://www.cime.unifi.it

CIME's activity is supported by:

– Istituto Nazionale di Alta Mathematica "F. Severi"
– Ministero degli Affari Esteri - Direzione Generale per la Promozione e la Cooperazione - Ufficio V
– Ministero dell'Istruzione, dell'Università e della Ricerca

Dan Abramovich · Marcos Mariño
Michael Thaddeus · Ravi Vakil

Enumerative Invariants in Algebraic Geometry and String Theory

Lectures given at the
C.I.M.E. Summer School
held in Cetraro, Italy
June 6–11, 2005

Editors:
Kai Behrend
Marco Manetti

 Springer

FONDAZIONE
CIME
ROBERTO CONTI

Authors and Editors

Dan Abramovich
Department of Mathematics
Box 1917
Brown University
Providence, RI 02912
USA
abrmovic@math.brown.edu

Kai Behrend
Department of Mathematics
University of British Columbia
1984 Mathematics Rd
Vancouver, BC, V6T 1Z2
Canada
behrend@math.ubc.ca

Marco Manetti
Department of Mathematics
"Guido Castelnuovo"
University of Rome "La Sapienza"
P.le Aldo Moro 5
00185 Rome
Italy
manetti@mat.uniroma1.it

Marcos Mariño
Department of Physics
Theory Division, CERN
University of Geneva
1211 Geneva
Switzerland
marcos@mail.cern.ch
Marcos.Marino.Beiras@cern.ch

Michael Thaddeus
Department of Mathematics
Columbia University
2990 Broadway
New York, NY 10027
USA
thaddeus@math.columbia.edu

Ravi Vakil
Department of Mathematics
Stanford University
Stanford, CA 94305–2125
USA
vakil@math.stanford.edu

ISBN 978-3-540-79813-2   ISBN 978-3-540-79814-9 (eBook)
DOI 10.1007/978-3-540-79814-9

Lecture Notes in Mathematics ISSN print edition: 0075-8434
ISSN electronic edition: 1617-9692

Library of Congress Control Number: 2008927358

Mathematics Subject Classification (2000): 14H10, 14H81, 14N35, 53D40, 81T30, 81T45

Printed on acid-free paper

9 8 7 6 5 4 3 2 1

springer.com

Preface

Starting with the middle of the 1980s, there has been a growing and fruitful interaction between algebraic geometry and certain areas of theoretical high-energy physics, especially the various versions of string theory. In particular, physical heuristics have provided inspiration for new mathematical definitions (such as that of Gromov–Witten invariants) leading in turn to the solution of (sometimes classical) problems in enumerative geometry. Conversely, the availability of mathematically rigorous definitions and theorems has benefitted the physics research by providing needed evidence, in fields where experimental testing seems still very far in the future.

This process is still ongoing in the present day, and actually expanding. A partial reflection of it can be found in the courses of the CIME session *Enumerative invariants in algebraic geometry and string theory*. The session took place in Cetraro from June 6 to June 11, 2005 with the following courses:

- Dan Abramovich (Brown University): *Gromov–Witten Invariants for Orbifolds.* (5 h)
- Marcos Mariño (CERN): *Open Strings.* (5 h)
- Michael Thaddeus (Columbia University): *Moduli of Sheaves.* (5 h)
- Ravi Vakil (Stanford University): *Gromov–Witten Theory and the Moduli Space of Curves.* (5 h)

Moreover, the following two talks were given as complementary material to the course of Abramovich.

- Jim Bryan (University of British Columbia): *Quantum cohomology of orbifolds and their crepant resolutions.*
- Barbara Fantechi (Sissa): *Virtual fundamental class.*

Orbifolds are a natural generalization of complex manifolds, where local charts are given not by open subsets of a complex vector space but by their quotients by finite groups. There are two natural descriptions, one in terms of charts (which actually works in symplectic geometry) and one in algebraic geometry as smooth complex Deligne–Mumford stacks. Physicists have long suggested treating orbifolds analogously to manifolds; this has led to the development of an orbifold Euler characteristic, then of orbifold Hodge numbers, and finally (due to Chen–Ruan, in 2000) of an orbifold cohomology, induced by a full theory of Gromov Witten invariants for orbifolds. The course of Dan Abramovich has presented the foundations, laid down in a series of papers by Abramovich, Vistoli and others, of the definition of Gromov–Witten invariants for orbifolds in the algebraic setting.

The course of Marcos Mariño presented explicit enumerative computations by manipulation of formal power series, based on the physical idea of transforming an open string theory to a closed string theory. The aim was to derive explicit relationships between Gromov–Witten, Donaldson–Thomas and Chern Simons invariants. In particular, the technique of the topological vertex was explained, which allows a cut-and-paste approach to the determination of such invariants. These methods have only partially received mathematical proofs; they are therefore an important source of conjectures and methods for further developments.

Enumerative geometry computations on moduli spaces of sheaves have long been extremely useful both in physics and in mathematics; for instance we may recall the definition of Donaldson and (more recently) Donaldson–Thomas invariants, for surfaces and (some) threefolds respectively. The course by Michael Thaddeus has been a very broad overview of this kind of techniques, with a particular accent on the definition of the Donaldson–Thomas invariants and the recent conjectures that relate them to Gromov–Witten invariants for Calabi Yau threefolds; evidence for the conjectures and examples illustrating their significance have also been included.

One of the more established parts of the algebraic geometry – high energy physics interaction has been the rigorous definition and the computation of Gromov–Witten invariants for smooth projective varieties. At the basis of the very definition there is the existence and properness of the moduli stack $\overline{M}_{g,n}$ of stable curves. Surprisingly, in recent years it has been possible to deduce theorems about $\overline{M}_{g,n}$ using the results of the Gromov–Witten theory. The course of Ravi Vakil gave a general introduction to this area of research, starting at a comparatively elementary level and then reaching proofs of some conjectures of C. Faber on the tautological cohomology ring of $\overline{M}_{g,n}$.

We acknowledge the COFIN 2003 "Spazi di moduli e teoria di Lie" for the partial financial support given to this C.I.M.E. session.

We express our deep gratitude to Barbara Fantechi, for her very active role in the organization, for help and for precious scientific advices to the second editor.

Kai Behrend
Marco Manetti

Contents

The Moduli Space of Curves and Gromov–Witten Theory

Lectures on Gromov–Witten Invariants of Orbifolds

D. Abramovich

Department of Mathematics, Box 1917, Brown University
Providence, RI 02912, USA
abrmovic@math.brown.edu

1 Introduction

1.1 What This Is

This text came out of my CIME minicourse at Cetraro, June 6–11, 2005. I kept the text relatively close to what actually happened in the course. In particular, because of last minutes changes in schedule, the lectures on usual Gromov–Witten theory started after I gave two lectures, so I decided to give a sort of introduction to non-orbifold Gromov–Witten theory, including an exposition of Kontsevich's formula for rational plane curves. From here the gradient of difficulty is relatively high, but I still hope different readers of rather spread-out backgrounds will get something out of it. I gave few computational examples at the end, partly because of lack of time. An additional lecture was given by Jim Bryan on his work on the crepant resolution conjecture with Graber and with Pandharipande, with what I find very exciting computations, and I make some comments on this in the last lecture. In a way, this is the original reason for the existence of orbifold Gromov–Witten theory.

1.2 Introspection

One of the organizers' not-so-secret reasons for inviting me to give these lectures was to push me and my collaborators to finish the paper [3]. The organizers were only partially successful: the paper, already years overdue, was only being circulated in rough form upon demand at the time of these lectures. Hopefully it will be made fully available by the time these notes are published. Whether they meant it or not, one of the outcomes of this intention of the organizers is that these lectures are centered completely around the paper [3]. I am not sure I did wisely by focussing so much on our work and not bringing in approaches and beautiful applications so many others have contributed, but this is the outcome and I hope it serves well enough.

K. Behrend, M. Manetti (eds.), *Enumerative Invariants in Algebraic Geometry and String Theory*. Lecture Notes in Mathematics 1947,
© Springer-Verlag Berlin Heidelberg 2008

1.3 Where Does All This Come From?

Gromov–Witten invariants of orbifolds first appeared, in response to string theory – in particular Zaslow's pioneering [41] – in the inspiring work of W. Chen and Y. Ruan (see, e.g. [15]). The special case of orbifold cohomology of finite quotient orbifolds was treated in a paper of Fantechi-Göttsche [20], and the special case of that for symmetric product orbifolds was also discovered by Uribe [39]. In the algebraic setting, the underlying construction of moduli spaces was undertaken in [4], [5], and the algebraic analogue of the work of Chen-Ruan was worked out in [2] and the forthcoming [3]. The case of finite quotient orbifolds was developed in work of Jarvis–Kauffmann–Kimura [25]. Among the many applications I note the work of Cadman [11], which has a theoretical importance later in these lectures.

In an amusing twist of events, Gromov–Witten invariants of orbifolds were destined to be introduced by Kontsevich but failed to do so – I tell the story in the appendix.

1.4 Acknowledgements

First I'd like to thank my collaborators Tom Graber and Angelo Vistoli, whose joint work is presented here, hopefully adequately. I thank the organizers of the summer school – Kai Behrend, Barbara Fantechi and Marco Manetti – for inviting me to give these lectures. Thanks to Barbara Fantechi and Damiano Fulghesu whose notes taken at Cetraro were of great value for the preparation of this text. Thanks are due to Maxim Kontsevich and Lev Borisov who gave permission to include their correspondence. Finally, it is a pleasure to acknowledge that this particular project was inspired by the aforementioned work of W. Chen and Ruan.

2 Gromov–Witten Theory

2.1 Kontsevich's Formula

Before talking of orbifolds, let us step back to the story of Gromov–Witten invariants. Of course these first came to be famous due to their role in mirror symmetry. But this failed to excite me, a narrow-minded algebraic geometer such as I am, until Kontsevich [28] gave his formula for the number of rational plane curves.

This is a piece of magic which I will not resist describing.

Setup

Fix an integer $d > 0$. Fix points $p_1, \ldots p_{3d-1}$ in general position in the plane. Look at the following number:

Definition 2.1.1.

$$N_d = \#\left\{\begin{array}{l} C \subset \mathbb{P}^2 \text{ a rational curve,} \\ \deg C = d, \text{ and} \\ p_1, \ldots p_{3d-1} \in C \end{array}\right\}.$$

Remark 2.1.1. *One sees that $3d - 1$ is the right number of points using an elementary dimension count: a degree d map of $\mathbb{P}^1$ to the plane is parametrized by three forms of degree d (with $3(d+1)$ parameters). Rescaling the forms (one parameter) and automorphisms of $\mathbb{P}^1$ (three parameters) should be crossed out, giving $3(d+1) - 1 - 3 = 3d - 1$.*

Statement

Theorem 2.1.1 (Kontsevich). *For $d > 1$ we have*

$$N_d = \sum_{\substack{d = d_1 + d_2 \\ d_1, d_2 > 0}} N_{d_1} N_{d_2} \left(d_1^2 d_2^2 \binom{3d-4}{3d_1-2} - d_1^3 d_2 \binom{3d-4}{3d_1-1} \right).$$

Remark 2.1.2. *The first few numbers are*

$$N_1 = 1, \ N_2 = 1, \ N_3 = 12, \ N_4 = 620, \ N_5 = 87304.$$

The first two are elementary, the third is classical, but N_4 and N_5 are nontrivial.

Remark 2.1.3. *The first nontrivial analogous number in $\mathbb{P}^3$ is the number of lined meeting four other lines in general position (the answer is 2, which is the beginning of Schubert calculus).*

2.2 Set-Up for a Streamlined Proof

$\overline{\mathcal{M}}_{0,4}$

We need one elementary moduli space: the compactified space of ordered four-tuples of points on a line, which we describe in the following unorthodox manner :

$$\overline{\mathcal{M}}_{0,4} = \overline{\left\{ p_1, p_2, q, r \in L \ \middle| \ \begin{array}{l} L \simeq \mathbb{P}^1 \\ p_1, p_2, q, r \text{ distinct} \end{array} \right\}}$$

The open set $\mathcal{M}_{0,4}$ indicated in the braces is isomorphic to $\mathbb{P}^1 \smallsetminus \{0, 1, \infty\}$, the coordinate corresponding to the cross ratio

$$CR(p_1, p_2, q, r) = \frac{p_1 - p_2}{p_1 - r} \frac{q - r}{q - p_2}.$$

The three points in the compactification, denoted

$$0 = (p_1, p_2 \mid q, r),$$
$$1 = (p_1, q \mid p_2, r), \quad \text{and}$$
$$\infty = (p_1, r \mid p_2, q),$$

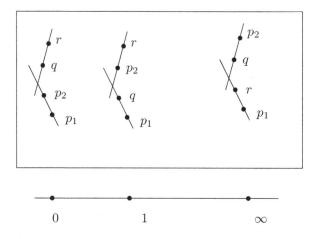

describe the three different ways to split the four points in two pairs and position them on a nodal curve with two rational components.

A One-Parameter Family

We now look at our points $p_1, \ldots p_{3d-1}$ in the plane.

We pass two lines ℓ_1, ℓ_2 with general slope through the last point p_{3d-1} and consider the following family of rational plane curves in $C \to B$ parametrized by a curve B:

- Each curve C_b contains $p_1, \ldots p_{3d-2}$ (but not necessarily p_{3d-1}).
- One point $q \in C_b \cap \ell_1$ is marked.
- One point $r \in C_b \cap \ell_2$ is marked.

In fact, we have a family of rational curves $C \to B$ parametrized by B, most of them smooth, but finitely many have a single node, and a morphism $f : C \to \mathbb{P}^2$ immersing the fibers in the plane.

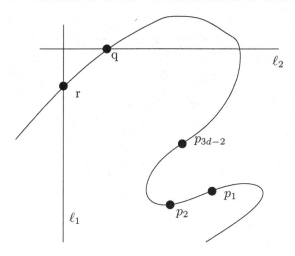

The Geometric Equation

We have a cross-ratio map

$$B \xrightarrow{\lambda} \overline{\mathcal{M}}_{0,4}$$

$$C \mapsto CR(p_1, p_2, q, r)$$

Since points on $\mathbb{P}^1$ are homologically equivalent we get

$$\deg_B \lambda^{-1}(p_1, p_2|q, r) = \deg_B \lambda^{-1}(p_1, q|p_2, r).$$

The Right-Hand Side

Now, each curve counted in $\deg_B \lambda^{-1}(p_1, q|p_2, r)$ is of the following form:

- It has two components C_1, C_2 of respective degrees d_1, d_2 satisfying $d_1 + d_2 = d$.
- We have $p_1 \in C_1$ as well as $3d_1 - 2$ other points among the $3d - 4$ points $p_3, \ldots p_{3d-2}$.
- We have $p_2 \in C_2$ as well as the remaining $3d_2 - 2$ points from $p_3, \ldots p_{3d-2}$.
- We select one point $z \in C_1 \cap C_2$ where the two abstract curves are attached.
- We mark one point $q \in C_1 \cap \ell_1$ and one point $r \in C_2 \cap \ell_2$.

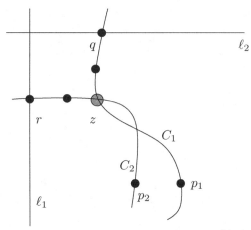

For every choice of splitting $d_1 + d_2 = d$ we have $\binom{3d-4}{3d_1-2}$ ways to choose the set of $3d_1 - 2$ points on C_1 from the $3d - 4$ points $p_3, \ldots p_{3d-2}$. We have N_{d_1} choices for the curve C_1 and N_{d_2} choices for C_2. We have $d_1 \cdot d_2$ choices for z, d_1 choices for q and d_2 for r. This gives the term

$$\deg_B \lambda^{-1}(p_1, q | p_2, r) = \sum_{\substack{d = d_1 + d_2 \\ d_1, d_2 > 0}} \binom{3d - 4}{3d_1 - 2} \cdot N_{d_1} N_{d_2} \cdot d_1 d_2 \cdot d_1 \cdot d_2.$$

A simple computation in deformation theory shows that each of these curves actually occurs in a fiber of the family $C \to B$, and it occurs exactly once with multiplicity 1.

The Left-Hand Side

Curves counted in $\deg_B \lambda^{-1}(p_1, p_2 | q, r)$ come in two flavors: there are *irreducible* curves passing through $q = r = \ell_1 \cap \ell_2$. This is precisely N_d.

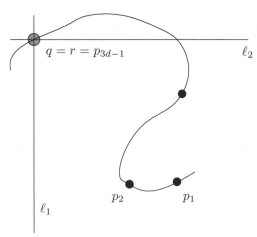

Now, each *reducible* curve counted in $\deg_B \lambda^{-1}(p_1, p_2 | q, r)$ is of the following form:

- It has two components C_1, C_2 of respective degrees d_1, d_2 satisfying $d_1 + d_2 = d$.
- We have $3d_1 - 1$ points among the $3d - 4$ points $p_3, \dots p_{3d-2}$ are on C_1.
- We have $p_1, p_2 \in C_2$ as well as the remaining $3d_2 - 2$ points from $p_3, \dots p_{3d-2}$.
- We select one point $z \in C_1 \cap C_2$. where the two abstract curves are attached.
- We mark one point $q \in C_1 \cap \ell_1$ and one point $r \in C_1 \cap \ell_2$.

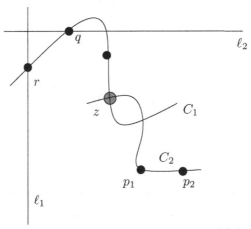

For every choice of splitting $d_1 + d_2 = d$ we have $\binom{3d-4}{3d_1-1}$ ways to choose the set of $3d_1 - 1$ points on C_1 from the $3d - 4$ points $p_3, \dots p_{3d-2}$. We have N_{d_1} choices for the curve C_1 and N_{d_2} choices for C_2. We have $d_1 \cdot d_2$ choices for z, d_1 choices for q and d_1 for r. This gives

$$\deg_B \lambda^{-1}(p_1, p_2 | q, r)$$

$$= N_d + \sum_{\substack{d = d_1 + d_2 \\ d_1, d_2 > 0}} \binom{3d - 4}{3d_1 - 1} \cdot N_{d_1} N_{d_2} \cdot d_1 d_2 \cdot d_1^2$$

Equating the two sides and rearranging we get the formula.   □

2.3 The Space of Stable Maps

Gromov–Witten theory allows one to systematically carry out the argument in general, without sweeping things under the rug as I have done above.

Kontsevich introduced the moduli space $\overline{\mathcal{M}}_{g,n}(X, \beta)$ of stable maps, a basic tool in Gromov–Witten theory. As it turned out later, it is a useful moduli space for other purposes.

Fixing a complex projective variety X, two integers $g, n \geq 0$ and $\beta \in H_2(X, \mathbb{Z})$, one defines:

$$\overline{\mathcal{M}}_{g,n}(X, \beta) = \{(f : C \to X, p_1, \ldots, p_n \in C)\},$$

where

- C is a nodal connected projective curve.
- $f : C \to X$ is a morphism with $f_*[C] = \beta$.
- p_i are n distinct points on the smooth locus of C.
- The group of automorphisms of f fixing all the p_i is finite.

Here an automorphism of f means an automorphism $\sigma : C \to C$ such that $f = f \circ \sigma$, namely a commutative diagram:

Remark 2.3.1. *if* $X = $ *a point, then* $\overline{\mathcal{M}}_{g,n}(X, 0) = \overline{\mathcal{M}}_{g,n}$, *the Deligne–Mumford stack of stable curves.*

Remark 2.3.2. *It is not too difficult to see that the stability condition on finiteness of automorphisms is equivalent to either of the following:*

- *The sheaf* $\omega_C(p_1 + \cdots + p_n) \otimes f^*M$ *is ample for any sufficiently ample sheaf* M *on* C.
- *We say that a point on the normalization of* C *is special if it is either a marked point or lies over a node of* C. *The condition is that any rational component* C_0 *of the normalization of* C *such that* $f(C_0)$ *is a point, has at least three special points, and any such elliptic component has at least one special point.*

The basic result, treated among other places in [30], [21], is

Theorem 2.3.1. $\overline{\mathcal{M}}_{g,n}(X, \beta)$ *is a proper Deligne–Mumford stack with projective coarse moduli space.*

2.4 Natural Maps

The moduli spaces come with a rich structure of maps tying them, and X, together.

Evaluation

First, for any $1 \leq i \leq n$ we have natural morphisms, called *evaluation morphisms*

$$\overline{\mathcal{M}}_{g,n}(X,\beta) \xrightarrow{\ e_i\ } X$$
$$(C \xrightarrow{f} X, p_1, \ldots, p_n) \mapsto f(p_i)$$

Contraction

Next, given a morphism $\phi : X \to Y$ and $n > m$ we get an induced morphism

$$\overline{\mathcal{M}}_{g,n}(X,\beta) \longrightarrow \overline{\mathcal{M}}_{g,m}(Y,\phi_*\beta)$$
$$(C \xrightarrow{f} X, p_1, \ldots, p_n) \mapsto \text{stabilization of } (C \xrightarrow{\phi \circ f} Y, p_1, \ldots, p_m).$$

Here in the stabilization we contract those rational components of C which are mapped to a point by $\phi \circ f$ and have fewer than three special points. This is well defined if either $\phi_*\beta \neq 0$ or $2g - 2 + n > 0$.

For instance, if $n > 3$ we get a morphism $\overline{\mathcal{M}}_{0,n}(X,\beta) \to \overline{\mathcal{M}}_{0,4}$.

2.5 Boundary of Moduli

Understanding the subspace of maps with degenerate source curve C is key to Gromov–Witten theory.

Fixed Degenerate Curve

Suppose we have a degenerate curve

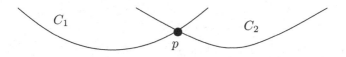

$$C = C_1 \overset{p}{\cup} C_2.$$

So C is a fibered coproduct of two curves. By the universal property of co-products

$$Hom(C,X) = Hom(C_1,X) \underset{Hom(p,X)}{\times} Hom(C_2,X)$$
$$= Hom(C_1,X) \underset{X}{\times} Hom(C_2,X)$$

Varying Degenerate Curve: The Boundary of Moduli

We can work this out in the fibers of the universal families. If we set $g = g_1 + g_2$, $n = n_1 + n_2$ and $\beta = \beta_1 + \beta_2$ we get a morphism

$$\overline{\mathcal{M}}_{g_1,n_1+1}(X,\beta_1) \times_X \overline{\mathcal{M}}_{g_2,n_2+1}(X,\beta_2) \longrightarrow \overline{\mathcal{M}}_{g,n}(X,\beta),$$

with the fibered product over e_{n_1+1} on the left and e_{n_2+1} on the right. On the level of points this is obtained by gluing curves C_1 at point $n_1 + 1$ with C_2 at point $n_2 + 1$ and matching the maps f_1, f_2. This is a finite unramified map, and we can think of the product on the left as a space of stable n-pointed maps of genus g and class β with a distinguished marked node.

Compatibility with Evaluation Maps

This map is automatically compatible with the other "unused" evaluation maps. For instance if $i \leq n_1$ we get a commutative diagram

$$
\begin{array}{ccc}
\overline{\mathcal{M}}_{g_1,n_1+1}(X,\beta_1) \underset{X}{\times} \overline{\mathcal{M}}_{g_2,n_2+1}(X,\beta_2) & \longrightarrow & \overline{\mathcal{M}}_{g,n}(X,\beta) \\
\pi_1 \downarrow & & \downarrow e_i \\
\overline{\mathcal{M}}_{g_1,n_1+1}(X,\beta_1) & \xrightarrow{\ e_i\ } & X.
\end{array}
$$

As these tend to get complicated we will give the markings their "individual labels" rather than a number (which may change through the gluing or contraction maps).

2.6 Gromov–Witten Classes

We start with some simplifying assumptions:

- $g = 0$.
- X is "convex" [21]: no map of rational curve to X is obstructed. Examples: $X = \mathbb{P}^r$, or any homogeneous space.

We simplify notation: $M = \overline{\mathcal{M}}_{0,n+1}(X,\beta)$. We take $\gamma_i \in H^*(X,\mathbb{Q})^{even}$ to avoid sign issues.

We now define *Gromov–Witten classes*:

Definition 2.6.1.

$$
\langle \gamma_1, \ldots, \gamma_n, * \rangle_\beta^X := e_{n+1 \, *}(e^*(\gamma_1 \times \ldots \times \gamma_n) \cap [M])) \in H^*(X).
$$

Here the notation is $e := e_1 \times \cdots \times e_n : M \to X^n$:

$$
\begin{array}{ccc}
M & \xrightarrow{\ e_{n+1}\ } & X \\
e := e_1 \times \cdots \times e_n \downarrow & & \\
X^n. & &
\end{array}
$$

When X is fixed we will suppress the superscript X from the notation.

2.7 The WDVV Equations

The main formula in genus 0 Gromov–Witten theory is the Witten–Dijkgraaf–Verlinde–Verlinde (or WDVV) formula. It is simplest to state for $n = 3$:

Theorem 2.7.1.

$$\sum_{\beta_1+\beta_2=\beta} \left\langle \langle \gamma_1, \gamma_2, * \rangle_{\beta_1}, \gamma_3, * \right\rangle_{\beta_2} = \sum_{\beta_1+\beta_2=\beta} \left\langle \langle \gamma_1, \gamma_3, * \rangle_{\beta_1}, \gamma_2, * \right\rangle_{\beta_2}.$$

For general $n \geq 3$ it is convenient to label the markings by a finite set I to avoid confusion with numbering. The formula is

$$\sum_{\beta_1+\beta_2=\beta} \sum_{A \sqcup B=I} \left\langle \langle \gamma_1, \gamma_2, \delta_{A_1}, \ldots, \delta_{A_k}, * \rangle_{\beta_1}, \gamma_3, \delta_{B_1}, \ldots, \delta_{B_m}, * \right\rangle_{\beta_2}$$

$$= \sum_{\beta_1+\beta_2=\beta} \sum_{A \sqcup B=I} \left\langle \langle \gamma_1, \gamma_3, \delta_{A_1}, \ldots, \delta_{A_k}, * \rangle_{\beta_1}, \gamma_2, \delta_{B_1}, \ldots, \delta_{B_m}, * \right\rangle_{\beta_2}$$

Note that the only thing changed between the last two lines is the placement of γ_2 and γ_3.

This formalism of Gromov–Witten classes and the WDVV equations is taken from a yet unpublished paper of Graber and Pandharipande. One advantage is that it works with cohomology replaced by Chow groups.

Gromov–Witten Numbers

The formalism that came to us from the physics world is equivalent, though different, and involves Gromov–Witten numbers, defined as follows:

Definition 2.7.1.

$$\langle \gamma_1, \ldots, \gamma_n \rangle_\beta^X := \int_M e^*(\gamma_1 \times \ldots \times \gamma_n),$$

where this time $M = \overline{\mathcal{M}}_{0,n}(X, \beta)$.

The following elementary lemma allows one to go back and forth between these formalisms:

Lemma 2.7.1.

1. $\langle \gamma_1, \ldots, \gamma_n, \gamma_{n+1} \rangle_\beta = \int_X \langle \gamma_1, \ldots, \gamma_n, * \rangle_\beta \cup \gamma_{n+1}$

2. Choose a basis $\{\alpha_i\}$ for $H^*(X, \mathbb{Q})$. Write the intersection matrix $\int_X \alpha_i \cup \alpha_j = g_{ij}$, and denote the inverse matrix entries by g^{ij}. Then

$$\langle \gamma_1, \ldots, \gamma_n, * \rangle_\beta = \sum_{i,j} \langle \gamma_1, \ldots, \gamma_n, \alpha_i \rangle_\beta \, g^{ij} \, \alpha_j.$$

The WDVV equations then take the form

$$\sum_{\beta_1+\beta_2=\beta} \sum_{A\sqcup B=I} \sum_{i,j} \langle \gamma_1,\gamma_2,\delta_A,\alpha_i\rangle_{\beta_1}\ g^{ij}\ \langle \alpha_j,\gamma_3,\delta_B,\gamma_4\rangle_{\beta_2}$$

$$= \sum_{\beta_1+\beta_2=\beta} \sum_{A\sqcup B=I} \sum_{i,j} \langle \gamma_1,\gamma_3,\delta_A,\alpha_i\rangle_{\beta_1}\ g^{ij}\ \langle \alpha_j,\gamma_2,\delta_B,\gamma_4\rangle_{\beta_2}$$

2.8 Proof of WDVV

We sketch the proof of WDVV, in the formalism of Gromov–Witten classes, under strong assumptions:

- $e_i : M \longrightarrow X$ is smooth.
- The contraction $M \longrightarrow \overline{\mathcal{M}}_{0,4}$ is smooth, and moreover each node can be smoothed out independently.

These assumptions hold for the so-called convex varieties discussed by Fulton and Pandharipande.

Setup

Now, it would be terribly confusing to use the usual numbering for the markings, the evaluation maps and the cohomology classes pulled back by the corresponding evaluation marks, as we will use the structure of the boundary discussed above. Instead we give them names.

The first moduli space we need is the space of genus 0 pointed stable maps to X with class β_1. The markings are used in three ways:

1. The first two are used to pull back γ_1, γ_2 on the left hand side (and γ_1, γ_3 on the right).
2. The next bunch is used to pull back the δ_{A_i}.
3. The last is used to push forward.

It is convenient to put together the first two sets of and call the result $\hat{A}$. The last marking will be denoted by the symbol ▶, suggesting something is to be glued on the right. For short notation we will use

$$\overline{\mathcal{M}}_1 = \overline{\mathcal{M}}_{0,\hat{A}\sqcup\blacktriangleright}(X,\beta_1).$$

We also use shorthand for the cohomology and homology classes:

$$\eta_1 = e_{\hat{A}}^*(\gamma_1 \times \gamma_2 \times \delta_A)$$
$$\xi_1 = \eta_1 \cap [M_1]$$

The second moduli space we need is the space of genus 0 pointed stable maps to X with class β_2. The markings are used in four ways:

1. The first will be used to pull back $\langle \gamma_1, \gamma_2, \delta_{A_1}, \ldots, \delta_{A_k}, * \rangle_{\beta_1}$. It is the marking used for gluing and is accordingly denoted ◄.
2. The next one is used to pull back γ_3 on the left hand side (and γ_2 on the right).
3. The next bunch is used to pull back the δ_{B_i}.
4. The last is used to push forward.

We leave the first one alone, we put together the next two sets of and call the result $\hat{B}$. The last marking will be denoted by the symbol •. The notation we will use is

$$\overline{\mathcal{M}}_2 = \overline{\mathcal{M}}_{0, \blacktriangleleft \sqcup \hat{B} \sqcup \bullet}(X, \beta_1).$$

The shorthand for the cohomology and homology classes is:

$$\eta_2 = e_{\hat{B}}^*(\gamma_3 \times \delta_B)$$
$$\xi_1 = \eta_2 \cap [M_2]$$

The key to the proof of WDVV is to express the class on either side in terms of something symmetric on the glued moduli space. We use the notation

$$\overline{\mathcal{M}}_1 \times_X \overline{\mathcal{M}}_1 \longrightarrow \overline{\mathcal{M}} := \overline{\mathcal{M}}_{0, \hat{A} \sqcup \hat{B} \sqcup \bullet}(X, \beta),$$

where the gluing is done with respect to $e_{\blacktriangleright} : \overline{\mathcal{M}}_1 \to X$ on the left, and $e_{\blacktriangleleft} : \overline{\mathcal{M}}_2 \to X$ on the right. A relevant symmetric class which comes up as a bridge between the two sides of the formula is

$$\eta_{12} = e_{\hat{A} \sqcup \hat{B}}^*(\gamma_1 \times \gamma_2 \times \delta_A \times \gamma_3 \times \delta_B).$$

The Fibered Product Diagram

Consider the diagram

$$
\begin{array}{ccccc}
\overline{\mathcal{M}}_1 \times_X \overline{\mathcal{M}}_2 & \xrightarrow{\ p_2\ } & \overline{\mathcal{M}}_2 & \xrightarrow{\ e_{\bullet}\ } & X \\
{\scriptstyle p_1} \downarrow & & \downarrow {\scriptstyle e_{\blacktriangleleft}} & & \\
\overline{\mathcal{M}}_1 & \xrightarrow{\ e_{\blacktriangleright}\ } & X. & &
\end{array}
$$

It is important that $e_{\blacktriangleleft}$ is smooth, in particular flat. We suppress Poincaré duality isomorphisms in the notation.

We now have by definition

$$\left\langle \langle \gamma_1, \gamma_2, \delta_A, * \rangle_{\beta_1}, \gamma_3, \delta_B, * \right\rangle_{\beta_2} = e_{\bullet *}\left(e_{\blacktriangleleft}^*(e_{\blacktriangleright *} \xi_1) \cap \xi_2 \right)$$

and the projection formula gives

$$= (e_{\bullet} \circ p_2)_*\left(p_1^* \eta_1 \cup p_2^* \eta_2 \cap [\overline{\mathcal{M}}_1 \times_X \overline{\mathcal{M}}_2] \right)$$

So we need to understand the class $p_1^* \eta_1 \cup p_2^* \eta_2 \cap [\overline{\mathcal{M}}_1 \times_X \overline{\mathcal{M}}_2]$ on the fibered product.

End of Proof

Since we have a big sum in the WDVV equation, we need to take the union of all these fibered products. They are put together by the gluing maps, as in the following fiber diagram:

$$
\begin{array}{ccc}
\coprod\limits_{\beta_1+\beta_2=\beta} \coprod\limits_{A \sqcup B=I} \overline{\mathcal{M}}_1 \times_X \overline{\mathcal{M}}_2 \xrightarrow{} \Delta^X_{(12|3\bullet)} \overset{\ell}{\hookrightarrow} \overline{\mathcal{M}} \\
\downarrow \qquad\qquad\qquad \downarrow{\scriptstyle st} \\
\{(12|3\bullet)\} \hookrightarrow \overline{\mathcal{M}}_{0,4}
\end{array}
$$

where the markings in $\overline{\mathcal{M}}_{0,4}$ are denoted $1, 2, 3$, and $\bullet$ to match with our other notation. The subscheme $\Delta^X_{(12|3\bullet)} \subset \overline{\mathcal{M}}$ is defined by this diagram.

The smoothness assumption on the stabilization/contraction map

$$
st : \overline{\mathcal{M}} \to \overline{\mathcal{M}}_{0,4}
$$

guarantees that the gluing map

$$
j : \coprod_{\beta_1+\beta_2=\beta} \coprod_{A \sqcup B=I} \overline{\mathcal{M}}_1 \times_X \overline{\mathcal{M}}_2 \to \Delta^X_{(12|3\bullet)}
$$

is finite and birational, and in fact we have an equality in Chow classes:

$$
\ell_* \left[\coprod_{\beta_1+\beta_2=\beta} \coprod_{A \sqcup B=I} \overline{\mathcal{M}}_1 \times_X \overline{\mathcal{M}}_2 \right] = st^* \left[\{(12|3\bullet)\} \right].
$$

We can now complete our equation:

$$
\sum\sum \Big\langle \langle \gamma_1, \gamma_2, \delta_A, * \rangle_{\beta_1}, \gamma_3, \delta_B, * \Big\rangle_{\beta_2}
$$
$$
= \sum\sum (e_\bullet \circ p_2)_* \left(p_1^* \eta_1 \cup p_2^* \eta_2 \cap [\overline{\mathcal{M}}_1 \times_X \overline{\mathcal{M}}_2] \right)
$$
$$
= \sum\sum (e_\bullet \circ p_2)_* \left(\ell^* \eta_{12} \cap [\overline{\mathcal{M}}_1 \times_X \overline{\mathcal{M}}_2] \right)
$$
$$
= e_{\bullet *} \left(\eta_{12} \cap st^* [\{(12|3\bullet)\}] \right),
$$

where the last evaluation map is $e_\bullet : \overline{\mathcal{M}} \to \overline{\mathcal{M}}_{0,4}$.

The latter expression is evidently symmetric and therefore the order of γ_2 and γ_3 is immaterial. The formula follows. $\qquad\qquad\square$

2.9 About the General Case

The WDVV equation holds in general when one replaces fundamental classes of moduli space in the smooth case with virtual fundamental classes, something that takes care of the non-smoothness in an organized manner. The drawback of virtual fundamental classes is that in general one loses the enumerative nature of Gromov–Witten classes. However the flexibility of the formalism allows one to compute cases where the classes are enumerative by going through cases where they are not.

I will definitely not try to get into the details here – they belong in a different lecture series. However I cannot ignore the subject completely – in the orbifold situation virtual fundamental classes are always necessary! Let me just indicate the principles.

First, the smoothness assumption of stabilization holds in the "universal case" – by which I mean to say that it holds for the moduli stacks of pre-stable curves introduced by Behrend [6]. We in fact have a fiber diagram extending the above:

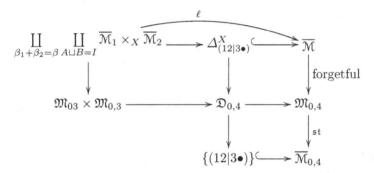

where $\mathfrak{st} : \mathfrak{M}_{0,4} \longrightarrow \overline{M}_{0,4}$ is flat. This is a situation where refined pull-backs can be used.

The formalism of algebraic virtual fundamental classes works for an arbitrary smooth projective X. It requires the definition of classes $[\overline{M}]^{\mathrm{vir}}$ in the Chow group of each moduli space $\overline{M}$. The key property that this satisfies is summarized as follows:

Consider the diagram

$$\overline{M}_1 \times \overline{M}_2 \longleftarrow \overline{M}_1 \times_X \overline{M}_2 \hookrightarrow \sqcup \overline{M}_1 \times_X \overline{M}_2 \longrightarrow \overline{M}$$
$$\downarrow \qquad\qquad\qquad \downarrow \qquad\qquad\qquad \downarrow \qquad\qquad\qquad \downarrow$$
$$X \times X \xleftarrow{\;\;\Delta\;\;} X \qquad\qquad \mathfrak{M}_{03} \times \mathfrak{M}_{0,3} \xrightarrow{\;\mathfrak{gl}\;} \mathfrak{M}_{0,4}.$$

The condition is:

$$\sum_{\beta_1+\beta_2=\beta} \sum_{A \sqcup B = I} \Delta^!\left([\overline{M}_1]^{\mathrm{vir}} \times [\overline{M}_2]^{\mathrm{vir}}\right) \;=\; \mathfrak{gl}^! [\overline{M}]^{\mathrm{vir}}.$$

For a complete algebraic treatment and an explanation why this is the necessary equation see [7].

The main theorem is

Theorem 2.9.1. *This equation holds for the class associated to obstruction theory*

$$[\overline{\mathcal{M}}]^{vir} \;=\; (\overline{\mathcal{M}}, E)$$

with

$$E = \mathbf{R}\pi_* f^* T_X$$

coming from the diagram of the universal curve

$$
\begin{array}{ccc}
\mathcal{C} & \xrightarrow{\;f\;} & X \\
{\scriptstyle \pi}\big\downarrow & & \\
\overline{\mathcal{M}} & &
\end{array}
$$

3 Orbifolds/Stacks

3.1 Geometric Orbifolds

One can spend an entire lecture series laying down the foundations of orbifolds or stacks. Since I do not have the luxury, I'll stick to a somewhat intuitive, and necessarily imprecise, presentation. The big drawback is that people who do not know the subject get only a taste of what it is about, and have to look things up to really understand what is going on. Apart from the standard references, one may consult the appendix of [40] or [18] for an introduction to the subject.

I will use the two words "orbifold" and "stack" almost interchangeably.

Geometrically, an orbifold X is locally given as a quotient of a space, or manifold, Y by the action of a finite group, giving a chart $[Y/G] \to X$. The key is to remember something about the action.

A good general way to do this is to think of X as the equivalence class of a groupoid

$$R \underset{s}{\overset{t}{\rightrightarrows}} V$$

where

- s, t are étale morphisms or, more generally, smooth morphisms.
- The morphism $s \times t : R \to V$ is required to be finite.

The notation used is $X \simeq [R \rightrightarrows V]$. You think of R as an equivalence relation on the points of V, and of X as the "space of equivalence classes".

This notation of a groupoid and its associated orbifold is very much a shorthand – the complete data requires a composition map

$$R \underset{{}_t V s}{\times} R \;\to\; R$$

as well as an identity morphism $V \to R$, satisfying standard axioms I will not write in detail. They are inspired by the case of a quotient:

A quotient is the orbifold associate to the following groupoid: $[V/G] = [R \rightrightarrows V]$, where $R = G \times V$, the source map $s : G \times V$ is the projection, and the target map $t : R \to V$ is the action.

I have not described the notion of equivalence of groupoids, neither did I describe the notion of a morphism of orbifolfds presented by groupoids. It is a rather complicated issue which I had rather avoid. The moduli discussion below will shed some light on it.

3.2 Moduli Stacks

Orbifolds come about rather frequently in the theory of moduli. In fact, the correct definition of an algebraic stack is as a sort of tautological solution to a moduli problem.

An algebraic stack $\mathcal{X}$ is by definition a category, and implicitly the category of families of those object we want to parametrize. It comes with a "structure functor" $\mathcal{X} \to Sch$ to the category of scheme, which to each family associates the base of the family. A morphism of stacks is a functor commuting with the structure functor.

Here is the key example: $\mathcal{M}_g$ – the moduli stack of curves. The category $\mathcal{M}_g$ has as its objects

$$\left\{ \begin{array}{c} \mathcal{C} \\ \downarrow \\ B \end{array} \right\}$$

where each $\mathcal{C} \to B$ is a family of curves of genus g. The structure functor $\mathcal{M}_g \to Sch$ of course sends the object $\mathcal{C} \to B$ to the base scheme B. It is instructive to understand what arrows we need. Of course we want to classify families up to isomorphisms, so isomorphisms

$$\begin{array}{ccc} \mathcal{C} & \overset{\sim}{\longrightarrow} & \mathcal{C}' \\ \downarrow & & \downarrow \\ B & = & B \end{array}$$

must be included. But it is not hard to fathom that pullbacks are important as well, and indeed a morphism in $\mathcal{M}_g$ is defined to be a *fiber diagram*

$$\begin{array}{ccc} \mathcal{C} & \longrightarrow & \mathcal{C}' \\ \downarrow & & \downarrow \\ B & \longrightarrow & B' \end{array}$$

That's why you'll hear the term "fibered category" used.

How is a scheme thought of as a special case of a stack? A scheme X is the moduli space of its own points. So an object is "a family of points of X parametrized by B", i.e. a morphism $B \to X$. So the stack associated to X is just the category of schemes over X.

Here is another example: consider a quotient orbifold $[Y/G]$, where Y is a variety and G a finite group. How do we think of it as a category? A point of the orbit space Y/G should be an orbit of G in Y, and you can think of an orbit as the image of an equivariant map of a principal homogeneous G-space P to Y. We make this the definition:

- An object of $[Y/G]$ over a base scheme B is a principal homogeneous G-space $P \to B$ together with a G-equivariant map $P \to Y$:

- An arrow is a fiber diagram:

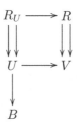

An important special case is when Y is a point, giving the classifying stack of G, denoted

$$\mathcal{B}G := [\{pt\}/G].$$

Objects are just principal G-bundles, and morphisms are fiber diagrams.

Kai Behrend gave an elegant description of the stack associated to a groupoid in general. First we note the following: for any scheme X and any étale surjective $V \to X$, one can write $R_V = V \times_X V$ and the two projections give a groupoid $R_V \rightrightarrows V$. If, as suggested above, R_V is to be considered as an equivalence relation on V, then clearly the equivalence classes are just points of X, so we had better define things so that $X = [R_V \rightrightarrows V]$.

Now given a general groupoid $R \rightrightarrows V$, an object over a base scheme B is very much like a principal homogeneous space: it consists of an étale covering $U \to B$, giving rise to $R_U \rightrightarrows U$ as above, together with maps $U \to V$ and $R_U \to R$ making the following diagram (and all its implicit siblings) *cartesian*:

There is an important object of $\mathfrak{X} = [R \rightrightarrows V]$ with the scheme V as its base: you take $U = R$ above, with the two maps $U \to B$ and $U \to V$ being the source and target maps $R \to V$ respectively. What it does is it gives an *étale covering* $V \to \mathfrak{X}$. The existence of such a thing is in fact an axiom required of a fibered category to be a Deligne–Mumford algebraic stack, but since I have not gotten into details you'll need to study this elsewhere. The requirement says in essence that every object should have a universal deformation space.

Some words you will see:

- An algebraic space is a stack of the form $[R \rightrightarrows V]$ with $R \to V \times V$ injective. This is where "stacks" meet "sheaves".
- An Artin stack, also known as a general algebraic stack, is what you get when you only require the source and target maps $R \to V$ to be smooth, and do not require $R \to V \times V$ to be proper either. You can't quite think of an Artin stack as "locally the quotient of a scheme by a group action" – the categorical viewpoint is necessary.

3.3 Where Do Stacks Come Up?

Moduli, Of Course

The first place where you meet stacks is when trying to build moduli spaces. Commonly, fine moduli spaces do not exist because objects have automorphisms, and stacks are the right replacement.

Hidden Smoothness

But even if you are not too excited by moduli spaces, stacks, in their incarnation as orbifolds, are here to stay. The reason is, it is often desirable to view varieties with finite quotient singularities as if they were smooth, and indeed to every such variety there is a relevant stack, which is indeed smooth.

This feature comes up, and increasing in appearance, in many topics in geometry: the minimal model program, mirror symmetry, geometry of three-manifolds, the McKay correspondence, and even in Haiman's $n!$ theorem (though this is not the way Haiman would present it).

3.4 Attributes of Orbifolds

If one is to study orbifolds along with varieties or manifolds, one would like to have tools similar to ones available for varieties and manifolds.

Indeed, the theory is well developed:

- Orbifolds have homology and cohomology groups, and cohomology of smooth orbifolds satisfies Poincaré duality with rational coefficients.

- Chow groups with rational coefficients for Deligne–Mumford stacks were constructed by Vistoli [40] and Gillet [22] independently in the 80s. More recently Kresch showed in his thesis that they have Chow groups with integer coefficients.
- One can talk about sheaves, K-theory and derived categories of stacks. That's a natural framework for the McKay correspondence.
- Smooth Deligne–Mumford stack have a dualizing invertible sheaf.
- Laumon and Moret-Bailly introduced the cotangent complex $\mathbb{L}_{\mathcal{X}}$ of an algebraic stack. This is rather easy for Deligne–Mumford stacks, but their construction was found to be flawed for Artin stack. This problem was recently corrected by Martin Olsson [34].
- Deligne–Mumford stacks have coarse moduli spaces – this is a theorem of Keel and Mori [27]. For $[Y/G]$ the moduli space is just the geometric quotient Y/G, namely the orbit space. In general this is an algebraic space X with a morphism $\mathcal{X} \to X$ which is universal, and moreover such that $\mathcal{X}(k)/\operatorname{Isom} \to X(k)$ is bijective whenever k is an algebraically closed field.
- The inertia stack: this is a natural stack associated to $\mathcal{X}$, which in a way points to where $\mathcal{X}$ fails to be a space. Every object ξ of $\mathcal{X}$ has its automorphism group $\operatorname{Aut}(\xi)$, and these can be put together in one stack $\mathcal{I}(\mathcal{X})$, whose objects are pairs (ξ, σ), with ξ an object of $\mathcal{X}$ and $\sigma \in \operatorname{Aut}(\xi)$. One needs to know that this is an algebraic stack when $\mathcal{X}$ is, a Deligne–Mumford stack when $\mathcal{X}$ is, etc. This follows from the abstract and not – too – illuminating formula

$$\mathcal{I}(\mathcal{X}) = \mathcal{X} \underset{\mathcal{X} \times \mathcal{X}}{\times} \mathcal{X},$$

where the product is taken relative to the diagonal map on both sides. As an example, if $\mathcal{X} = \mathcal{B}G$ then $\mathcal{I}(\mathcal{X}) = [G/G]$, where G acts on itself by conjugation.

There is one feature which one likes to ignore, but there comes a point where one needs to face the facts of life: stacks are not a category! They are a 2-category: arrows are functors, 2-arrows are natural transformations.

3.5 Étale Gerbes

This is a class of stacks which will come up in our constructions.

Informally, an étale gerbe is a stack which locally (in the étale topology) looks like $X \times \mathcal{B}G$, with G a finite group.

Formally (but maybe not so intuitively), it is a Deligne–Mumford stack $\mathcal{X}$ such that the morphisms $\mathcal{I}(\mathcal{X}) \to \mathcal{X} \to X$ are all finite étale.

We need to tie this thing better with a group G. We will only need to consider the case where G is abelian.

An étale gerbe $\mathcal{G} \to X$ is said to be *banded* by the finite abelian group G if one is given, for every object $\xi \in \mathcal{G}(S)$, an isomorphism $G(S) \simeq \operatorname{Aut}(\xi)$ in a functorial manner.

Gerbes banded by G can be thought of as principal homogeneous spaces under the "group stack" $\mathcal{B}G$. As such, they are classified by the "next cohomology group over", $H^2_{\text{ét}}(X, G)$. This led Giraud in his thesis under Grothendieck to define the non-abelian second cohomology groups using non-abelian gerbes, but this goes too far afield for us.

4 Twisted Stable Maps

4.1 Stable Maps to a Stack

Consider a semistable elliptic surface, with base B and a section. We can naturally view this as a map $B \to \overline{M}_{1,1}$. Angelo Vistoli, when he was on sabbatical at Harvard in 1996, asked the following beautiful, and to me very inspiring, question: what's a good way to compactify the moduli of elliptic surfaces? can one use stable maps to get a good compactification?

Now consider in general:

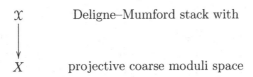

In analogy to $\overline{M}_{g,n}(X, \beta)$, we want a compact moduli space of maps $C \to \mathcal{X}$.

One can define stable maps as in the scheme case, but there is a problem: the result is not compact. As Angelo Vistoli likes to put it, trying to work with a non-compact moduli space is like trying to keep your coins when you have holes in your pockets. The solution that comes naturally is that

the source curve $\mathcal{C}$ must acquire a stack structure as well as it degenerates!

Both problem and solution are clearly present in the following example, which is "universal" in the sense that we take $\mathcal{X}$ to be a one parameter family of curves itself:

Consider $\mathbb{P}^1 \times \mathbb{P}^1$ with coordinates x, s near the origin and the projection with coordinate s onto $\mathbb{P}^1$. Blowing up the origin we get a family of curves, with general fiber $\mathbb{P}^1$ and special fiber a nodal curve, with local equation $xy = t$ at the node. Taking base change $\mathbb{P}^1 \to \mathbb{P}^1$ of degree 2 with equation $t^2 = s$ we get a singular scheme X with a map $X \to \mathbb{P}^1$ given by coordinate s. This is again a family of $\mathbb{P}^1$s with nodal special fiber, but local equation $xy = s^2$.

This is a quotient singularity, and using the chart $[\mathbb{A}^2/(\mathbb{Z}/2\mathbb{Z})]$ with coordinates u, v satisfying $u^2 = x, v^2 = y$ we get a smooth orbifold $\mathcal{X}$, with coarse moduli space X and a map $\mathcal{X} \to \mathbb{P}^1$. It is a family of $\mathbb{P}^1$s parametrized by $\mathbb{P}^1$, degenerating to an orbifold curve.

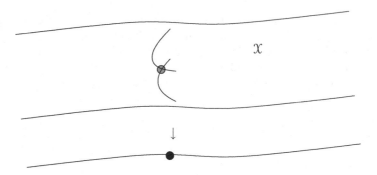

If you think about the family of stable maps $\mathbb{P}^1 \to X$ parametrized by $\mathbb{P}^1 \smallsetminus \{0\}$ given by the embedding of $\mathbb{P}^1$ in the corresponding fiber, there simply isn't any stable map from a nodal curve that can be fit over the missing point $\{0\}$! The only reasonable thing to fit in there is the fiber itself, which is an orbifold nodal curve. We call these *twisted curves*.

4.2 Twisted Curves

This is what happens in general: degenerations force us to allow stacky (or twisted) structure at the nodes. Thinking ahead about gluing curves we see that we had better allow these structures at markings as well.

A *twisted curve* is a gadget as follows:

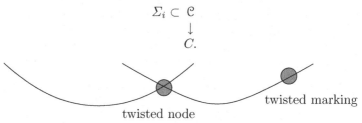

Here

- C is a nodal curve.
- $\mathcal{C}$ is a Deligne–Mumford stack with C as its coarse moduli space.
- Over a node $xy = 0$ of C, the twisted curve $\mathcal{C}$ has a chart

$$[\{uv = 0\}/\boldsymbol{\mu}_r]$$

where the action of the cyclotomic group $\boldsymbol{\mu}_r$ is described by

$$(u, v) \mapsto (\zeta u, \zeta^{-1} v).$$

We call this kind of action, with two inverse weights ζ, ζ^{-1}, a *balanced action*. It is necessary for the existence of smoothing of $\mathcal{C}$! In this chart, the map $\mathcal{C} \to C$ is given by $x = u^r, y = v^r$.

- At a marking, $\mathcal{C}$ has a chart $[\mathbb{A}^1/\mu_r]$, with standard action $u \mapsto \zeta u$, and the map is $x = u^r$.
- The substack Σ_i at the i-th marking is locally defined by $u = 0$. This stack Σ_i is canonically an étale gerbe banded by μ_r.

Note that we introduce stacky structure only at isolated points of C and never on whole components. Had we added stack structures along components, we would get in an essential manner a 2-stack, and I don't really know how to handle these.

As defined, twisted curves form a 2-category, but it is not too hard to show it is equivalent to a category, so we are on safe grounds.

The automorphism group of a twisted curve is a fascinating object – I'll revisit it later.

This notion of twisted curves was developed in [5]. As we discovered later, a similar idea appeared in Ekedahl's [19].

4.3 Twisted Stable Maps

Definition 4.3.1. *A twisted stable map consists of*

$$(f : \mathcal{C} \to \mathcal{X}, \Sigma_1, \ldots, \Sigma_n),$$

where

- $\Sigma_i \subset \mathcal{C}$ *gives a pointed twisted curve.*
- $\mathcal{C} \xrightarrow{f} \mathcal{X}$ *is a representable morphism.*
- *The automorphism group* $\mathrm{Aut}_{\mathcal{X}}(f, \Sigma_i)$ *of f fixing Σ_i is finite.*

I need to say something about the last two *stability conditions*, necessary for the moduli problem being separated.

Representability of $f : \mathcal{C} \to \mathcal{X}$ means that for any point x of $\mathcal{C}$ the associated map

$$\mathrm{Aut}(x) \to \mathrm{Aut}(f(x))$$

on automorphisms is injective. So the orbifold structure on $\mathcal{C}$ is the "most economical" possible, in that we do not add unnecessary automorphisms.

The second condition is in analogy with the usual stable map case, and indeed it can be replaced by conditions on ampleness of a suitable sheaf or number of special points on rational and elliptic components. Most conveniently, it is equivalent to the following schematic condition: the map of coarse moduli spaces

$$f : C \to X$$

is stable.

But as I defined things I have not told you what an element of $\mathrm{Aut}_{\mathcal{X}}(f, \Sigma_i)$ is! In fact, to make this into a stack I need a category of families of such twisted stable maps.

Definition 4.3.2. *A map from* $(f : \mathcal{C} \to \mathcal{X}, \Sigma_1, \ldots, \Sigma_n)$ *over* S *to* $(f' : \mathcal{C}' \to \mathcal{X}', \Sigma'_1, \ldots, \Sigma'_n)$ *over* S' *is the following:*

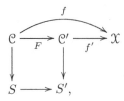

consisting of

- *A fiber diagram with morphism F as above*
- *A 2-isomorphism $\alpha : f \to f' \circ F$*

Note that the notion of automorphisms is more subtle than the case of stable maps to a scheme, even if $\mathcal{C}$ is a scheme. For instance, in the case $\mathcal{X} = \overline{\mathcal{M}}_g$, a map $\mathcal{C} \to \mathcal{X}$ is equivalent to a fibered surface $S \to C$ with fibers of genus g, and S can easily have automorphisms acting on the fibers and keeping C fixed, for instance if the fibers are hyperelliptic!

4.4 Transparency 25: The Stack of Twisted Stable Maps

The Stack of Twisted Stable Maps

The first nontrivial fact that we have here is the following: the collection of twisted stable maps is again a 2-category, simply because twisted curves are naturally a 2-subcategory of the 2-category of stacks. But there is a simple lemma saying that every 2-morphism between 1-morphisms is unique and invertible when it exists. This is precisely the condition guaranteeing the following:

Fact. The 2-category of twisted stable maps is equivalent to a category.

Here we come to a sticky point: This category is a generalization of $\overline{\mathcal{M}}_{g,n}(X, \beta)$. But in our applications we wish to sometimes insert moduli spaces denoted $\overline{\mathcal{M}}$ in there for $\mathcal{X}$ – our original application was $\mathcal{X} = \overline{\mathcal{M}}_{1,1}$! We made a decision to avoid confusion and denote the category

$$\mathcal{K}_{g,n}(\mathcal{X}, \beta),$$

after Kontsevich. Some people have objected quite vocally, but I think the choice is sound and the objections are not too convincing and a bit too late for us. But you are perfectly welcome to use notation of your choice.

The main result is:

Theorem 4.4.1. *The category $\mathcal{K}_{g,n}(\mathcal{X}, \beta)$ is a proper Deligne–Mumford stack with projective coarse moduli space.*

Here β is simply the class of the curve $f_*[C]$ on the coarse moduli space X.

On the Proof

To prove this, Vistoli and I had to go through several chambers of hell. As far as I can see, the complete symplectic proof (using Fukaya-Ono) is not easier.

Our main difficulty was the fact that a basic tool like Hilbert schemes was not available for our construction. Now a much better approach, due almost entirely to Martin Olsson, is available. I will sketch is now, because I think it is beautiful. Readers who are not keen on subtleties of constructing moduli stacks might prefer to skip this and take the theorem entirely on faith.

Olsson's proof has the following components:

- He first constructs rather explicitly the stack of twisted curves with its universal family [36]:

$$\mathfrak{C}^{\mathrm{tw}}_{g,n}$$
$$\downarrow$$
$$\mathfrak{M}^{\mathrm{tw}}_{g,n}.$$

- In great generality he constructs [35] a stack of morphisms between two given stacks, and identifies the substack

$$\mathcal{K}_{g,n}(\mathcal{X}, \beta) \quad \subset \quad \mathrm{Hom}_{\mathfrak{M}^{\mathrm{tw}}_{g,n}}(\mathfrak{C}^{\mathrm{tw}}_{g,n}, \mathcal{X}).$$

- He further shows in the same paper that when passing to coarse moduli spaces, the natural morphism

$$\mathrm{Hom}_{\mathfrak{M}^{\mathrm{tw}}_{g,n}}(\mathfrak{C}^{\mathrm{tw}}_{g,n}, X) \quad \longrightarrow \quad \mathrm{Hom}_{\mathfrak{M}^{\mathrm{tw}}_{g,n}}(\mathfrak{C}_{g,n}, X)$$

is of finite type, implying the same for $\mathcal{K}_{g,n}(\mathcal{X}, \beta) \to \overline{\mathcal{M}}_{g,n}(X, \beta)$.

- To prove properness one can use the valuative criterion, whose proof in [5] is appropriate. In the same paper one counts and sees that $\mathcal{K}_{g,n}(\mathcal{X}, \beta) \to \overline{\mathcal{M}}_{g,n}(X, \beta)$ has finite fibers, implying projectivity.

4.5 Twisted Curves and Roots

Martin Olsson constructs the stack of twisted curves in general using logarithmic structures [36]. This is a very nice construction, but it would have been too much to introduce yet another big theory in these lectures. What I want to do here is describe a variant of this using root stacks, which works nicely in the case of tree-like curves – i.e. where the dual graph is a tree, equivalently every node separates the curve in two connected components.

The construction was first invented by Angelo Vistoli, but his treatment (see [3], Appendix B) has not yet appeared in print. I lectured on this at ICTP, but did not include in the lecture notes. Charles Cadman discovered this construction independently and used it to great advantage in his thesis [11], where a treatment is published.

Definition 4.5.1. *Consider a scheme X, a line bundle L, a section $s \in \Gamma(X, L)$, and a positive integer r. Define a stack*

$$\sqrt[r]{(L/X, s)}$$

whose objects over a scheme Y are $(f : Y \to X, M, \phi, t)$ where

- *M is a line bundle on Y and $t \in \Gamma(Y, M)$.*
- *$\phi : M^{\otimes r} \xrightarrow{\sim} L$.*
- *$\phi(t^r) = s$.*

Arrows are fiber diagrams as usual.

For a Cartier divisor D, Vistoli uses the notation

$$\sqrt[r]{(X, D)} := \sqrt[r]{(\mathcal{O}_X(D)/X, \mathbf{1}_D)}.$$

Cadman uses the notation $X_{D,r}$.

This stack $\sqrt[r]{(X, D)}$ or $X_{D,r}$ is isomorphic to X away from the zero set D of the section, and canonically introduces a stack structure with index r along D, which is "minimal" if D is smooth. This immediately enables us to define the stacky structure of a twisted curve at a marking starting with the coarse curve:

$$(C, p) \quad \rightsquigarrow \quad \mathcal{C} = \sqrt[r]{(C, p)} = C_{p,r}.$$

The case of a node is more subtle, and is best treated universally. Here we need to assume that the nodes are separating to use root stacks directly, otherwise one needs either subtle descent or logarithmic structures.

Assume given:

- A versal deformation space of nodal curves $C \to V$, with V a polydisk or a strictly henselian scheme
- $D \subset V$ the smooth divisor where a particular node in the fibers is preserved
- $Z \subset C$ the locus of these nodes, assumed separating
- $E_1, E_2 \subset C$ the two connected components of the preimage of D separated by Z

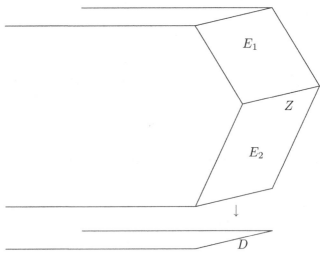

We have the structure morphism $V \to \mathfrak{M} := \mathfrak{M}_{g,n}$. Denote by $\mathfrak{M}_r^{\text{tw}}$ the locus in $\mathfrak{M}_{g,n}^{\text{tw}}$ where the given node is given stacky structure of index r, and $\mathfrak{C}_r^{\text{tw}}$ the universal twisted curve. Then we have

$$V \underset{\mathfrak{M}}{\times} \mathfrak{M}_r^{\text{tw}} = \sqrt[r]{(V, D)}$$

$$V \underset{\mathfrak{M}}{\times} \mathfrak{C}_r^{\text{tw}} = \sqrt[r]{(C, E_1)} \underset{C}{\times} \sqrt[r]{(C, E_2)}.$$

One can pore over these formulas for a long time to understand them. One thing I like to harvest from the first formula is a description of automorphisms: since $\mathfrak{M}_r^{\text{tw}} \to \mathfrak{M}$ is birational, but the versal deformation is branched with index r over D, this branching is accounted for by automorphisms of the twisted curve. We deduce that the automorphism group of a twisted curve fixing C is

$$\operatorname{Aut}_C(\mathfrak{C}) = \prod_{s \in \operatorname{Sing} \mathfrak{C}} \Gamma_s,$$

where $\Gamma_s \simeq \mu_{r_s}$ is the stabilizer of the corresponding node.

These automorphisms acting trivially on C are completely absent from the simple-minded orbifold picture, as in [15]. Alessio Corti calls them *ghost automorphisms*, and their understanding is a key to the paper [1].

4.6 Valuative Criterion for Properness

Vistoli's Purity Lemma

The key ingredient in proving the valuative criterion for properness is the following lemma, stated and proven by Vistoli:

Lemma 4.6.1 (Vistoli). *Consider the following commutative diagram:*

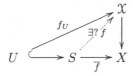

where

- $\mathcal{X}$ *is a Deligne–Mumford stack and* $\pi : \mathcal{X} \to X$ *is the coarse moduli space map.*
- S *is a smooth variety and* $U \subset S$ *is open with compliment of codimension* ≥ 2.

Then there exists $f : S \to \mathcal{X}$ *making the diagram commutative, unique up to a unique isomorphism.*

I love this lemma (and its proof). It seems that the lemma loves me back, as it has carried me through half my career!

Related ideas appeared in Mochizuki's [33].

Proof of the Lemma

For the proof, we may assume (working analytically) that S is a 2-disc and U is a punctured 2-disc. Hence U is simply connected. We can make $\mathcal{X}$ smaller as well and assume that $\mathcal{X} = [V/G]$ where G is a finite group fixing the origin, chosen so that its image in X is the image via $\bar{f}$ of $S \smallsetminus U = p$.

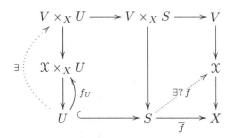

Note that $V \times_X U \to \mathcal{X} \times_X U$ is étale and proper, hence the section $U \to \mathcal{X} \times_X U$ lifts to $V \times_X U$ because U is simply connected. Let $\bar{U}$ be the closure of the image of U in $V \times_X S$.

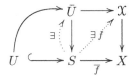

We obtain by projection $\bar{U} \to S$ which is finite birational and since S is smooth it has a section. Composing with $\bar{U} \to \mathcal{X}$ yields f.

Proof of the Valuative Criterion

Let me sketch how the valuative criterion for properness of $\mathcal{K}_{g,n}(\mathcal{X}, \beta)$ follows. Consider a punctured smooth curve $V \subset B$ and a family of stable maps

$$
\begin{array}{ccc}
C_V & \xrightarrow{f} & \mathcal{X} \\
\downarrow & & \\
V. & &
\end{array}
$$

By properness of $\overline{\mathcal{M}}_{g,n}(X, \beta)$ we may assume this extends to a family of stable maps

$$
\begin{array}{ccc}
C & \xrightarrow{f} & X \\
\downarrow & & \\
B. & &
\end{array}
$$

Using properness of $\mathfrak{X}$ and base change (and abhyankar's lemma on fundamental groups – characteristic 0 is essential!) we may assume that the map $C \to X$ lifts to $C \dashrightarrow \mathfrak{X}$ defined on a neighborhood of the generic point of every component of the fiber C_0. So it is defined on an open set $U \subset C$ whose complement has codimension 2. The purity lemma says that this map extends on the regular locus of C.

We are left to deal with singular points, where C is described locally as $xy = t^r$ for some r. The purity lemma says this extends over the local universal cover given by $uv = t$, where $u^r = x$ and $v^r = y$, and the fundamental group of the punctured neighborhood is cyclic of order r. Of course r changes when you do base change, but there is a smallest r' such that the map extends over the cover given by $u^{r'} = x, v^{r'} = y$, and this one is representable.

5 Gromov–Witten Classes

5.1 Contractions

Many of the features of the stacks of stable maps are still true for twisted stable maps. For instance, of one has a morphism $f : \mathfrak{X} \to \mathcal{Y}$ and $m < n$, then as long as either $2g - 2 + m > 0$ or $f_*\beta \neq 0$ we have a canonical map

$$\mathcal{K}_{g,n}(\mathfrak{X}, \beta) \to \mathcal{K}_{g,m}(\mathcal{Y}, f_\beta),$$

the construction of which is just a bit more subtle than the classical stabilization procedure.

5.2 Gluing and Rigidified Inertia

Much more subtle is the issue of gluing, and the related evaluation maps. To understand it we consider a nodal twisted curve with a separating node:

$$\mathcal{C} = \mathcal{C}_1 \overset{\Sigma}{\sqcup} \mathcal{C}_2.$$

As expected, one can prove that $\mathcal{C}$ is a coproduct in a suitable stack-theoretic sense, and therefore

$$\mathrm{Hom}(\mathcal{C}, \mathfrak{X}) \quad = \quad \mathrm{Hom}(\mathcal{C}_1, \mathfrak{X}) \underset{\mathrm{Hom}(\Sigma, \mathfrak{X})}{\times} \mathrm{Hom}(\mathcal{C}_2, \mathfrak{X}).$$

but Σ is no longer a point but a gerbe! We must ask:

- How can we understand $\mathrm{Hom}(\Sigma, \mathfrak{X})$?
- What is the universal picture?

Since Σ is a gerbe banded by $\boldsymbol{\mu}_r$, the nature of $\mathrm{Hom}(\Sigma, \mathfrak{X})$ definitely depends on r.

Definition 5.2.1. *Define a category*

$$\bar{\mathcal{I}}(\mathcal{X}) \;\; = \;\; \coprod_r \bar{\mathcal{I}}_r(\mathcal{X}),$$

where each component has objects

$$\bar{\mathcal{I}}_r(\mathcal{X})(T) \;\; = \;\; \left\{ \begin{array}{c} \mathcal{G} \xrightarrow{\phi} \mathcal{X} \\ \downarrow \\ T \end{array} \right\}$$

where

- $\mathcal{G} \to T$ *is a gerbe banded by μ_r.*
- $\phi : \mathcal{G} \to \mathcal{X}$ *is representable.*

A priori this is again a 2-category, but again and for a different reason, it is equivalent to a category. In fact we have

Theorem 5.2.1. $\bar{\mathcal{I}}(\mathcal{X})$ *is a Deligne–Mumford stack.*

There is a close relationship between $\bar{\mathcal{I}}(\mathcal{X})$ and the inertia stack $\mathcal{I}(\mathcal{X})$. In fact it is not too difficult to see that there is a diagram

$$\begin{array}{ccc} \mathcal{I}(\mathcal{X}) & \longrightarrow & \mathcal{X} \\ \downarrow & & \\ \bar{\mathcal{I}}(\mathcal{X}) & & \end{array}$$

making $\mathcal{I}(\mathcal{X})$ the universal gerbe over $\bar{\mathcal{I}}(\mathcal{X})$! The stack $\bar{\mathcal{I}}_r(\mathcal{X})$ can be constructed as the rigidification of the order r part

$$\mathcal{I}_r(\mathcal{X}) = \{(\xi, g)| g \in \mathrm{Aut}\xi, g \text{ of order } r\}$$

of the inertia stack by removing the action of the cyclic group $\langle g \rangle$ of-order r from the picture. This is analogous to the construction of Picard scheme, where the $\mathbb{C}^*$ automorphisms are removed from the picture. This is discussed in [1], [3] and in [37], and the notation is

$$\bar{\mathcal{I}}_r(\mathcal{X}) = \mathcal{I}_r(\mathcal{X}) \,\!\!\fatslash\, \mu_r.$$

In a 2-categorical sense, this is the same as saying that the group-stack $\mathcal{B}\mu_r$ acts 2-freely on $\mathcal{I}_r(\mathcal{X})$ and in fact

$$\bar{\mathcal{I}}_r(\mathcal{X}) = \mathcal{I}_r(\mathcal{X})/\mathcal{B}\mu_r,$$

but this tends to make me dizzy.

The name we give $\bar{\mathcal{I}}(\mathcal{X})$ is *the rigidified inertia stack.*

The following example may be illuminating: consider the global quotient stack $\mathcal{X} = [Y/G]$. Then a simple analysis shows:

$$\mathfrak{I}(\mathcal{X}) \;=\; \coprod_{(g)} \left[\, Y^g \,\big/\, C(g) \,\right],$$

where the union is over conjugacy classes (g) and $C(g)$ denotes the centralizer of g. But by definition the cyclic group $\langle g \rangle$ acts trivially on Y^g, therefore the action of $C(g)$ factors through

$$\overline{C(g)} := C(g)/\langle g \rangle.$$

We have

$$\bar{\mathfrak{I}}(\mathcal{X}) \;=\; \coprod_{(g)} \left[\, Y^g \,\big/\, \overline{C(g)} \,\right].$$

5.3 Evaluation Maps

We now have a natural evaluation map

$$\mathcal{K}_{g,n}(\mathcal{X}, \beta) \xrightarrow{\;e_i\;} \bar{\mathfrak{I}}(\mathcal{X})$$

$$(f : \mathcal{C} \to \mathcal{X}, \Sigma_1, \ldots \Sigma_n) \;\mapsto\; \Sigma_i \xrightarrow{\;f_{\Sigma_i}\;} \mathcal{X}$$

One point needs to be clarified: the gerbe Σ_i needs to be banded in order to give an object of $\bar{\mathfrak{I}}(\mathcal{X})$. This is automatic – as $\boldsymbol{\mu}_r$ canonically acts on the tangent space of $\mathcal{C}$ at Σ_i, this gives a canonical identification of the automorphism group of a point on Σ_i with $\boldsymbol{\mu}_r$!

There is a subtle feature we have to add here: when gluing curves we need to make the glued curve balanced. Therefore on one branch the banding has to be inverse to the other. We wire this into the definitions as follows. There is a natural involution

$$\mathfrak{I}(\mathcal{X}) \xrightarrow{\;\iota\;} \mathfrak{I}(\mathcal{X})$$

$$(\xi, \sigma) \mapsto (\xi, \sigma^{-1})$$

which gives rise to an involution, denoted by the same symbol, $\iota : \bar{\mathfrak{I}}(\mathcal{X}) \to \bar{\mathfrak{I}}(\mathcal{X})$.

On the level of gerbes, it sends $\mathcal{G} \to \mathcal{X}$ to itself, but for an object y of $\mathcal{G}$ changes the isomorphism $\boldsymbol{\mu}_r \simeq \mathrm{Aut}(Y)$ by composing with the homomorphism $\boldsymbol{\mu}_r \to \boldsymbol{\mu}_r$ sending $\zeta_r \mapsto \zeta_r^{-1}$.

We define a new *twisted evaluation map*

$$\check{e}_i = \iota \circ e_i : \mathcal{K}_{g,n}(\mathcal{X}, \beta) \longrightarrow \bar{\mathfrak{I}}(\mathcal{X}).$$

5.4 The Boundary of Moduli

We can finally answer the tow questions asked before:

Given a nodal curve $\mathcal{C} = \mathcal{C}_1 \overset{\Sigma}{\sqcup} \mathcal{C}_2$, we have

$$\mathrm{Hom}(\mathcal{C}, \mathcal{X}) \;=\; \mathrm{Hom}(\mathcal{C}_1, \mathcal{X}) \underset{\bar{\mathcal{I}}(\mathcal{X})}{\times} \mathrm{Hom}(\mathcal{C}_2, \mathcal{X}).$$

where the fibered product is with respect to $\check{e}_{\blacktriangleright} : \mathcal{C}_1 \to \bar{\mathcal{I}}(\mathcal{X})$ on the left and $e_{\blacktriangleleft} : \mathcal{C}_2 \to \bar{\mathcal{I}}(\mathcal{X})$ on the right.

We can apply this principle on universal curves, and obtain, just as in the case of usual stable maps, and obtain a morphism

$$\mathcal{K}_{g_1, n_1 + \blacktriangleright}(\mathcal{X}, \beta_1) \underset{\bar{\mathcal{I}}(\mathcal{X})}{\times} \mathcal{K}_{g_2, n_2 + \blacktriangleleft}(\mathcal{X}, \beta_2) \longrightarrow \mathcal{K}_{g,n}(X, \beta),$$

with the fibered product over the twisted evaluation map $\check{e}_{\blacktriangleright}$ on the left and the non-twisted $e_{\blacktriangleleft}$ on the right. Again this is compatible with the other evaluations, e.g. for $i \leq n_1$ we have a commutative diagram

$$
\begin{array}{ccc}
\mathcal{K}_{g_1, n_1 + \blacktriangleright}(\mathcal{X}, \beta_1) \underset{\bar{\mathcal{I}}(\mathcal{X})}{\times} \mathcal{K}_{g_2, n_2 + \blacktriangleleft}(\mathcal{X}, \beta_2) & \longrightarrow & \mathcal{K}_{g,n}(\mathcal{X}, \beta) \\
{\scriptstyle \pi_1} \downarrow & & \downarrow {\scriptstyle e_i} \\
\mathcal{K}_{g_1, n_1 + \blacktriangleright}(\mathcal{X}, \beta_1) & \xrightarrow{\;\; e_i \;\;} & \bar{\mathcal{I}}(\mathcal{X}) \,.
\end{array}
$$

We now come to a central observation of orbifold Gromov–Witten theory:

Since evaluation maps lie in $\bar{\mathcal{I}}(\mathcal{X})$, Gromov–Witten classes operate on the cohomology of $\bar{\mathcal{I}}(\mathcal{X})$, and not of $\mathcal{X}$!!

Of course $\mathcal{X}$ is part of the picture:

$$\mathcal{X} = \bar{\mathcal{I}}_1(\mathcal{X}) \subset \bar{\mathcal{I}}(\mathcal{X}).$$

The other pieces of $\bar{\mathcal{I}}(\mathcal{X})$ are known as *twisted sectors* (yet another meaning of "twisting" in geometry), and arise in various places in mathematics and physics for different reasons. But from the point of view of Gromov–Witten theory, the reason they arise is just the observation above, which comes down to the fact that a nodal twisted curve is glued along a gerbe, not a point.

5.5 Orbifold Gromov–Witten Classes

First, an easy technicality: we have a locally constant function

$$r : \bar{\mathcal{I}}(\mathcal{X}) \longrightarrow \mathbb{Z}$$

defined by sending

$$\bar{\mathfrak{I}}_r(\mathfrak{X}) \mapsto r$$

On the level of the non-rigidified inertia stack $\mathfrak{I}(\mathfrak{X})$, it simply means that

$$(\xi, \sigma) \quad \mapsto \quad \text{the order of } \sigma.$$

A locally constant function gives an element of $H^0(\mathfrak{I}(\mathfrak{X}), \mathbb{Z})$, and therefore we can multiply any cohomology class by this r.

Now, as observed above, Gromov–Witten theory operates on $H^*(\bar{\mathfrak{I}}(\mathfrak{X}))$. Because of its role in Gromov–Witten theory, the cohomology space $H^*(\bar{\mathfrak{I}}(\mathfrak{X}))$ got a special name – it is commonly known as the orbifold cohomology of $\mathfrak{X}$, denoted by

$$H^*_{orb}(\mathfrak{X}) := H^*(\bar{\mathfrak{I}}(\mathfrak{X})).$$

We again simplify notation: $\mathfrak{K} = \mathfrak{K}_{0,n+1}(\mathfrak{X}, \beta)$, and take $\gamma_i \in H^*_{orb}(\mathfrak{X})^{even}$ to avoid sign issues.

We now define *Gromov–Witten classes*:

Definition 5.5.1.

$$\langle \gamma_1, \ldots, \gamma_n, * \rangle^{\mathfrak{X}}_{\beta} := r \cdot \check{e}_{n+1 *}(e^*(\gamma_1 \cup \ldots \cup \gamma_n) \cap [\mathfrak{K}]^{vir})) \in H^*_{orb}(\mathfrak{X}).$$

I'm going to claim that WDVV works for these classes just as before, but any reasonable person will object – this factor of r must enter somewhere, and I had better explain why it is there and how it works out!

There are two ways I want to answer this. First – on a formal level, this factor r is needed because of the following: consider the stack $\mathfrak{D}(12|34)^{tw} \to \mathfrak{M}^{tw}_{0,4}$ consisting of twisted nodal curves where markings numbered $1, 2$ are separated from $3, 4$ by a node. This is not the fibered product $(\mathfrak{M}_{0,3} \times \mathfrak{M}_{0,3}) \times_{\mathfrak{M}_{0,4}} \mathfrak{M}^{tw}_{0,4}$, and there is exactly a multiplicity r involved – I explain this below. (In [2] we used a slightly different formalism, replacing the moduli stacks $\mathfrak{K}_{g,n}(\mathfrak{X}, \beta)$ with the universal gerbes, and a different correction is necessary.)

The second answer is important at least from a practical point of view. The Gromov–Witten theory of $\mathfrak{X}$ wants to behave as if there is a lifted evaluation map $\tilde{e}_i : \mathfrak{K}_{g,n}(\mathfrak{X}, \beta) \to \mathfrak{I}(\mathfrak{X})$, lifting $e_i : \mathfrak{K}_{g,n}(\mathfrak{X}, \beta) \to \bar{\mathfrak{I}}(\mathfrak{X})$.

$$
\begin{array}{ccc}
& & \mathfrak{I}(\mathfrak{X}) \\
& \overset{\exists \tilde{e}_i}{\nearrow} & \big\downarrow{\pi} \\
\mathfrak{K}_{g,n}(\mathfrak{X}, \beta) & \underset{e_i}{\longrightarrow} & \bar{\mathfrak{I}}(\mathfrak{X})
\end{array}
$$

Of course pulling back gives a multiplicative isomorphism

$$\pi^* : H^*(\bar{\mathfrak{I}}(\mathfrak{X}), \mathbb{Q}) \to H^*(\mathfrak{I}(\mathfrak{X}), \mathbb{Q}),$$

but a cohomological lifting of, say $\tilde{e}_{i *}$ of $e_{i *}$ is obtained rather by composing with the non-multiplicative isomorphism

$$(\pi_*)^{-1} : H^*(\bar{\mathfrak{I}}(\mathfrak{X}), \mathbb{Q}) \to H^*(\mathfrak{I}(\mathfrak{X}), \mathbb{Q}),$$

and $(\pi_*)^{-1} = r \cdot \pi^*$, so

$$\tilde{e}_{i*} = (\pi_*)^{-1} \circ e_{i*} = r\pi^* \circ e_{i*}.$$

Similarly we define

$$\tilde{e}_i^* = e_i^* \circ (\pi^*)^{-1}.$$

Since $(\pi^*)^{-1} = r \cdot \pi_*$ we can also write $\tilde{e}_i^* = r \cdot e_i^* \circ \pi_*$.[1] The factor r is cancelled out beautifully in WDVV, and one has all the formula as in the classical case. I'll indicate how this works in an example below.

I must admit that I am not entirely satisfied with this situation. I wish there were some true map standing for $\tilde{e}_i$. This might be very simple (why not replace $\bar{\mathfrak{I}}(\mathfrak{X})$ by $\bar{\mathfrak{I}}(\mathfrak{X}) \times \mathcal{B}\boldsymbol{\mu}_r$?), but the 2-categorical issues make my head spin when I think about it.

I also need to say something $[\mathfrak{X}]^{\mathrm{vir}}$.

5.6 Fundamental Classes

Let us see now where the differences and similarities are in the statement and formal proof.

The virtual fundamental class is again defined by the relative obstruction theory

$$E = \mathbf{R}\pi_* f^* T_{\mathfrak{X}}$$

coming from the diagram of the universal curve

$$
\begin{array}{ccc}
\mathcal{C} & \overset{f}{\longrightarrow} & \mathfrak{X} \\
{\scriptstyle \pi}\downarrow & & \\
\mathfrak{K}. & &
\end{array}
$$

It satisfies the Behrend–Fantechi relationship

$$\sum_{\beta_1+\beta_2=\beta} \sum_{A\sqcup B=I} \Delta^! \left([\mathfrak{K}_1]^{\mathrm{vir}} \times [\mathfrak{K}_2]^{\mathrm{vir}}\right) = \mathfrak{gl}^! [\mathfrak{K}]^{\mathrm{vir}}.$$

as in the diagram

$$
\begin{array}{ccccc}
\mathfrak{K}_1 \times \mathfrak{K}_2 & \longleftarrow \mathfrak{K}_1 \times_{\bar{\mathfrak{I}}(\mathfrak{X})} \mathfrak{K}_2 \longhookrightarrow & \sqcup \mathfrak{K}_1 \times_{\bar{\mathfrak{I}}(\mathfrak{X})} \mathfrak{K}_2 & \longrightarrow & \mathfrak{K} \\
\downarrow & \downarrow & \downarrow & & \downarrow \\
\bar{\mathfrak{I}}(\mathfrak{X}) \times \bar{\mathfrak{I}}(\mathfrak{X}) & \overset{\Delta}{\longleftarrow} \bar{\mathfrak{I}}(\mathfrak{X}) & \mathfrak{D}^{\mathrm{tw}}(12|3\bullet) & \overset{\mathfrak{gl}}{\longrightarrow} & \mathfrak{M}_{0,4}.
\end{array}
$$

---

[1] Notice that the latter formula was written incorrectly in [2], Sect. 4.5. We are indebted to Charles Cadman for noting this error. He also used these "cohomological evaluation maps" very effectively in his work.

The proof of this relationship is almost the same as that of Behrend–Fantechi, with one added ingredient: in comparing E with $E_i = \mathbf{R}\pi_{i\,*} f_i^* T_{\mathcal{X}}$ coming from

$$
\begin{array}{ccc}
\mathcal{C}_i & \xrightarrow{\ f\ } & \mathcal{X} \\
\pi \downarrow & & \\
\mathcal{K}_i, & &
\end{array}
$$

the "difference" is accounted for by $\pi_{\Sigma\,*} f_\Sigma^* T_{\mathcal{X}}$ as in

$$
\begin{array}{ccc}
\Sigma & \xrightarrow{\ f\ } & \mathcal{X} \\
\pi \downarrow & & \\
\mathcal{K}_1 \times_{\bar{\mathcal{I}}(\mathcal{X})} \mathcal{K}_2. & &
\end{array}
$$

The crucial step in proving the basic relationship is the following "tangent bundle lemma":

Lemma 5.6.1. *Assume* $e : S \to \bar{\mathcal{I}}(\mathcal{X})$ *corresponds to*

$$
\begin{array}{ccc}
\mathcal{G} & \xrightarrow{\ f\ } & \mathcal{X} \\
\pi \downarrow & & \\
S. & &
\end{array}
$$

Then there is a canonical isomorphism

$$
e^* T_{\bar{\mathcal{I}}(\mathcal{X})} \xrightarrow{\ \sim\ } \pi_* f^* T_{\mathcal{X}}.
$$

I will not prove the lemma here, but I wish to avow that it is a wonderful proof (which almost fits in the margins).

6 WDVV, Grading and Computations

6.1 The Formula

The WDVV formula says again:

Theorem 6.1.1.

$$
\sum_{\beta_1+\beta_2=\beta} \sum_{A \sqcup B = I} \left\langle \langle \gamma_1, \gamma_2, \delta_{A_1}, \ldots, \delta_{A_k}, * \rangle_{\beta_1}, \gamma_3, \delta_{B_1}, \ldots, \delta_{B_m}, * \right\rangle_{\beta_2}
$$

$$
= \sum_{\beta_1+\beta_2=\beta} \sum_{A \sqcup B = I} \left\langle \langle \gamma_1, \gamma_3, \delta_{A_1}, \ldots, \delta_{A_k}, * \rangle_{\beta_1}, \gamma_2, \delta_{B_1}, \ldots, \delta_{B_m}, * \right\rangle_{\beta_2}
$$

The fibered product diagram looks like this:

$$\mathcal{K}_1 \times_{\bar{\mathcal{J}}(\mathcal{X})} \mathcal{K}_2 \xrightarrow{p_2} \mathcal{K}_2 \xrightarrow{\check{e}_\bullet} \bar{\mathcal{J}}(\mathcal{X})$$

with vertical maps p_1 and $e_\blacktriangleleft$, and bottom row

$$\mathcal{K}_1 \xrightarrow{\check{e}_\blacktriangleright} \bar{\mathcal{J}}(\mathcal{X}).$$

We now have by definition

$$\left\langle \langle \gamma_1, \gamma_2, \delta_A, * \rangle_{\beta_1}, \gamma_3, \delta_B, * \right\rangle_{\beta_2} = r\check{e}_{\bullet *}\left(e_\blacktriangleleft^* \left(r\check{e}_{\blacktriangleright *} \xi_1 \right) \cap \xi_2 \right)$$

and the projection formula gives

$$= (r\check{e}_\bullet \circ p_2)_* \left(rp_1^* \eta_1 \cup p_2^* \eta_2 \cap \Delta^! [\mathcal{K}_1 \times \mathcal{K}_2] \right)$$

where the r inside the parentheses is pulled back from the glued markings. The overall factor r is the same on all terms of the equation so it can be crossed out in the proof.

The real difference comes in the divisor diagram, which becomes the following fiber diagram:

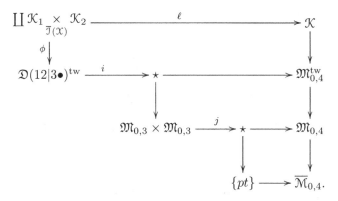

Now the map j is still birational, but i has degree $1/r$, which exactly cancels the factor r we introduced in the definition of Gromov–Witten classes.

6.2 Quantum Cohomology and Its Grading

Small Quantum Cohomology

In order to make explicit calculations with Gromov–Witten classes it is useful to have a guiding formalism. Consider, for instance the *small quantum cohomology product*. Let $N^+(X)$ be the monoid of homology classes of effective curves on X. Consider the formal monoid-algebra $\mathbb{Q}[[N^+(X)]]$ in which we represent a generator corresponding to $\beta \in N^+(X)$ multiplicatively as $\mathbf{q}^\beta$. Define a bilinear map on $QH^*(\mathcal{X}) := H^*(\bar{\mathcal{J}}(\mathcal{X}), \mathbb{Q}) \otimes \mathbb{Q}[[N^+(X)]]$ by the following rule on generators $\gamma_i \in H^*(\bar{\mathcal{J}}(\mathcal{X}), \mathbb{Q})$:

$$\gamma_1 * \gamma_2 = \sum_{\beta \in N^+(X)} \langle \gamma_1, \gamma_2, * \rangle_\beta \cdot \mathbf{q}^\beta.$$

This is a skew-commutative ring and as in the non-orbifold case, WDVV gives its associativity.

A somewhat simpler ring, but still exotic, is obtained by setting all $\mathbf{q}^\beta = 0$ for nonzero β. The product becomes

$$\gamma_1 \cdot \gamma_2 = \langle \gamma_1, \gamma_2, * \rangle_0.$$

This is the so-called *orbifold cohomology ring* (or *string cohomology*, according to Kontsevich). We denote it $H^*_{orb}(\mathcal{X})$. The underlying group is $H^*(\bar{\mathcal{J}}(\mathcal{X}), \mathbb{Q})$.

But one very very useful fact is that these are graded rings, and the grading, discovered by physicists, is fascinating.

Age of a Representation

A homomorphism $\rho : \boldsymbol{\mu}_r \to \mathbb{G}_m$ is determined by an integer $0 \le k \le r - 1$, as $\rho(\zeta_r) = \zeta_r^a$. We define

$$age(\rho) := k/r.$$

This extends by linearity to a function on the representation ring $age : R\boldsymbol{\mu}_r \to \mathbb{Q}$.

Age of a Gerbe in $\mathcal{X}$

Now consider a morphism $e : T \to \bar{\mathcal{J}}(\mathcal{X})$ corresponding to

$$\begin{array}{ccc} \mathcal{G} & \xrightarrow{f} & \mathcal{X} \\ \downarrow & & \\ T. & & \end{array}$$

The pullback $f^*T_{\mathcal{X}}$ is a representation of $\boldsymbol{\mu}_r$, and we can define $age(e) := age(T_{\mathcal{X}})$.

We can thus define a locally constant function $age : \bar{\mathcal{J}}(\mathcal{X}) \to \mathbb{Q}$, the value on any component being the age of any object e evaluating in this component.

Orbifold Riemann–Roch

There are numerous justifications for the definition of age in the literature, some look a bit like voodoo. In fact, the true natural justification for the definition of age is Riemann–Roch on orbifold curves.

Consider a twisted curve $\mathcal{C}$ and a locally free sheaf $\mathcal{E}$ on $\mathcal{C}$. There is a well defined notion of degree $\deg \mathcal{E} \in \mathbb{Q}$. It can be defined by the property that if D is a curve and $\phi : D \to \mathcal{C}$ is surjective of degree k on each component, then $\deg_{\mathcal{C}} \mathcal{E} = \deg_D \phi^* \mathcal{E}/k$

Riemann–Roch says the following:

Theorem 6.2.1.

$$\chi(\mathcal{C}, \mathcal{E}) \;=\; \mathrm{rank}(\mathcal{E}) \cdot \chi(\mathcal{C}, \mathcal{O}_{\mathcal{C}}) \;+\; \deg \mathcal{E} \;-\; \sum_{p_i \ marked} age_{p_i} \mathcal{E}$$

This is easy to show for a smooth twisted curve, and not very difficult for a nodal one as well.

6.3 Grading the Rings

We now define
$$H^i_{orb}(\mathcal{X}) \;=\; \bigoplus_{\Omega} H^{i \,-\, 2\,age(\Omega)}(\Omega, \mathbb{Q}),$$

where the sum is over connected components $\Omega \subset \bar{\mathcal{I}}(\mathcal{X})$.

We also define
$$\deg \mathbf{q}^{\beta} \;=\; 2\, \beta' \cdot c_1(T_{\mathcal{X}})$$

where β' is any class satisfying $\pi_* \beta' = \beta$.

Here is the result:

Theorem 6.3.1.

1. $H^*_{orb}(\mathcal{X})$ is a graded ring.
2. The product on $QH^*(\mathcal{X})$ is homogeneous, so it is a "pro-graded" ring.

6.4 Examples

Example 6.4.1. *Consider* $\mathcal{X} = \mathcal{B}G$. *In this case*

$$\mathcal{I}(\mathcal{X}) \;=\; [G/G] \;=\; \coprod_{(g)} \mathcal{B}(C(g))$$

where G acts on itself by conjugation. This case is simple as there are no nonzero curve classes and $QH^*(\mathcal{X}) = H^*_{orb}(\mathcal{X}) = \oplus_{(g)} \mathbb{Q}$. Since the tangent bundle is zero the ages are all 0, and all obstruction bundles vanish – i.e. the *virtual fundamental class equals the fundamental class. Also, since $\mathcal{I}(X) \to \bar{\mathcal{I}}(X)$ has a section, we have lifted evaluation maps $\tilde{e}_i : \mathcal{K}_{0,3}(\mathcal{B}G, 0) \to \mathcal{I}(X)$.*

Let us identify the components of $\mathcal{K}_{0,3}(\mathcal{B}G, 0)$:

There is one component $\mathcal{K}_{(g,h)}$ for each conjugacy class of triple $g, h, (gh)^{-1}$, describing the monodromy of a G-covering of $\mathbb{P}^1$ branched at $0, 1, \infty$. The automorphism group of such a cover is $C(g) \cap C(h)$, thus $\mathcal{K}_{(g,h)} \simeq \mathcal{B}(C(g) \cap C(h))$. The third (twisted, lifted) evaluation map is

$$\check{e}_3 : \mathcal{K}_{g,h} \to \mathcal{B}(C(gh))$$

of degree

$$\deg \check{e}_3 = \frac{|C(gh)|}{|C(g) \cap C(h)|}.$$

Therefore we get

$$x_{(g)} \cdot x_{(h)} = \sum_{(g,h)} \frac{|C(gh)|}{|C(g) \cap C(h)|} x_{(gh)}$$

where the sum runs over all simultaneous conjugacy classes of the pair (g, h). In other words, $QH^*_{orb}(BG) = Z(\mathbb{Q}[G])$, the center of the group ring of G.

Example 6.4.2. *Maybe the simplest nontrivial case is the following, but already here we can see some of the subtleties of the subject. I'll give these in full detail here – once you get the hang of it, it gets pretty fast!*

We take $X = \mathbb{P}(p, 1)$ where p is a prime. It is known as the teardrop projective line, a name inspired by imagining this to be a Riemann sphere with a "slightly pinched" north pole, with conformal angle $2\pi/p$.

It is defined to be the quotient of $\mathbb{C}^2 \setminus \{0\}$ by the $\mathbb{C}^*$ action $\lambda(x, y) = (\lambda^p x, \lambda y)$. In this case $H_2(X)$ is cyclic with a unique positive generator β_1, and writing $\mathbf{q} = q^{\beta_1}$ we have $\deg \mathbf{q} = 2(p+1)/p$ (in general for the weighted projective space $\mathbb{P}(a, b)$, with a, b coprime we have $\deg \mathbf{q} = 2(1/a + 1/b)$).

We can describe the inertia stack as follows:

$$\mathcal{I}(X) = X \sqcup \coprod_{i=1}^{p-1} \mathcal{B}(\mathbb{Z}/p\mathbb{Z}).$$

The rigidified inertia stack is therefore

$$\bar{\mathcal{I}}(\mathfrak{X}) = \mathfrak{X} \sqcup \coprod_{i=1}^{p-1} \Omega_i,$$

where the rigidified twisted sectors Ω_i are just points. Since Ω_i corresponds to the element $i \in \mathbb{Z}/p\mathbb{Z}$, and i acts on the tangent space of $\mathfrak{X}$ via ζ_p^i, its age is i/p.

Choose as generator of $H^2(\mathfrak{X})$ the class $x = c_1(\mathcal{O}_{\mathfrak{X}}(1))$; we have $\deg x = 1/p$. Choose also generators A_i to be each a generator of $H^0(\Omega_i)$ for $1 \leq i \leq p-1$. Since the age is i/p, it is positioned in degree $2i/p$ in $H^*_{orb}(\mathfrak{X})$. We conclude that the orbifold cohomology group, with grading, looks as follows:

$$\begin{array}{ccccccc} \text{degree } 0 & 2/p & \cdots & 2(p-1)/p & 2 \\ \mathbb{Q} \oplus & \mathbb{Q}A_1 \oplus & \cdots \oplus & \mathbb{Q}A_{p-1} & \oplus \mathbb{Q}x \end{array}$$

This is a good case to work out the difference between working with e_i : $\mathcal{K}_{0,3}(\mathfrak{X}, \beta) \to \bar{\mathcal{I}}(\mathfrak{X})$ and the lifted $\tilde{e}_i : \mathcal{K}_{0,3}(\mathfrak{X}, \beta) \to \mathcal{I}(\mathfrak{X})$ (which still exists on the relevant components of $\mathcal{K}_{0,3}(\mathfrak{X}, \beta)$ in this example).

Consider for instance the product $A_i * A_1$ when $i < p - 1$. In order to match the degrees, the result must be a multiple of A_{i+1}, and the only curve class β possible is 0. The component of $\mathcal{K}_{0,3}(\mathfrak{X}, 0)$ evaluating at $i, 1$ and the inverse $p - i - 1$ of $i + 1$ is, just as in the first example, the classifying stack of the joint centralizer of 1 and i, which is just $\mathcal{B}(\mathbb{Z}/p\mathbb{Z})$. Since A_1 and $A - i$ are fundamental classes, the pullback $e_1^* A_i \cup e_2^* A_1$ is the fundamental class.

But how about the virtual fundamental class? Here we use a fundamental fact about the virtual fundamental class: if it a multiple of the fundamental class, then it coincides with the fundamental class. We calculate

$$\check{e}_{3*}(e_1^* A_i \cup e_2^* A_1) = 1/pA_{i+1},$$

since the degree of $\check{e}_3$ is $1/p$. Since $r = p$ we get

$$A_i * A_1 = A_i \cdot A_1 = A_{i+1}.$$

Note that, had we used $\tilde{e}_i$ instead of e_i and replaced A_i by $\tilde{A}_i = \pi^* A_i$, the change in the degrees of the maps e_i would cancel out with the factor r and we would get the same formula:

$$\tilde{A}_i * \tilde{A}_1 = \tilde{A}_i \cdot \tilde{A}_1 = \tilde{A}_{i+1}.$$

Back to the calculation. By associativity, it follows that $A_i * A_j = A_i \cdot A_j = A_{i+j}$ as long as $i + j < p$.

The next product to calculate is $A_1 * A_{p-1}$. The result must be a multiple of x, again the only class β involved is $\beta = 0$. Again the virtual fundamental class is the fundamental class, and we have

$$\check{e}_{3\,*}(e_1^* A_1 \cup e_2^* A_{p-1}) = x,$$

as it has degree $1/p$. This time $r = 1$ so

$$A_1 * A_{p-1} = A_1 \cdot A_{p-1} = x.$$

The interesting product is $x * A_1$. It is easy to see that $x \cdot A_1 = 0$ because there is nothing in degree $2 + 2/p$ in $H^*_{orb}(\mathcal{X})$, and the only possible contribution to $x * A_1$ comes with $\beta = \beta_1$ of degree $2(1 + 1/p)$. What do we see in $\mathcal{K}_{0,3}(\mathcal{X}, \beta_1)$? Maps evaluating in Ω_1, parametrize twisted curves generically of the form $\mathcal{X}$, with one (in our case, the second) marking over the orbifold point and two freely roaming around $\mathcal{X}$. If we represent x by $1/p \cdot [z]$, with z one of the non-stacky points, then the class $e_1^* x \cup e_1^* A_1$ is represented by a family of twisted stable maps parametrized by the position of the third marking – that is a copy of $\mathcal{X}$! The restriction of the virtual fundamental class to the locus $e_1^{-1} z$ is again the fundamental class, and $e_3 = \check{e}_3$ is an isomorphism. We get $[z] * A_1 = \mathbf{q}$, or $x * A_1 = p^{-1}\mathbf{q}$. which completely determines the ring as

$$QH^*(\mathcal{X}) = \mathbb{Q}[[\mathbf{q}]][A_1]/(pA_1^{p+1} - \mathbf{q}).$$

Again, had we used $\tilde{e}_i$ instead of e_i and replaced A_i by $\tilde{A}_i = \pi^* A_i$, the change in the degrees of the maps e_i would cancel out with the factor r and we would get the same ring $\mathbb{Q}[[\mathbf{q}]][\tilde{A}_1]/(p\tilde{A}_1^{p+1} - \mathbf{q})$. This is Cadman's preferred formalism.

The more general case of $\mathbb{P}(a, b)$ is included in [3].

6.5 Other Work

Already in their original papers, Chen and Ruan gave a good number of examples of orbifold cohomology rings. Much work has been done on orbifold cohomology since, and I have no chance of doing justice to all the contributions. I'll mention a few which come to my rather incomplete memory: Fantechi and Göttsche [20] introduced a method for calculating orbifold cohomology of global quotients and figured out key examples; Uribe [39] also studied symmetric powers; Borisov, Chen and Smith [8] described the orbifold cohomology of toric stacks; Jarvis, Kaufmann and Kimura [26] and Goldin, Holm and Knutson [24] following Chen and Hu [14] described orbifold cohomology directly without recourse to Gromov–Witten theory.

Computations in Gromov–Witten theory of stacks beyond orbifold cohomology are not as numerous. C. Cadman [11, 12] computed Gromov–Witten invariants of $\sqrt[r]{(\mathbb{P}^2, C)}$ with C a smooth cubic, and derived the number of rational plane curves of degree d with tangency conditions to the cubic C. H.-H. Tseng [38] generalized the work of Givental and Coates on the mirror predictions to the case of orbifolds, a subject I'll comment on next.

6.6 Mirror Symmetry and the Crepant Resolution Conjecture

When string theorists first came up with the idea of mirror symmetry [13], the basic example was that of the quintic threefold Q and its mirror X, which happens to have Gorenstein quotient singularities. Mirror symmetry equates period integrals on the moduli space of complex structures on X with Gromov–Witten numbers of Q, and vice versa. String theorists had no problem dealing with orbifolds [16], [17], but they also had the insight of replacing X, or the stack $\mathcal{X}$, by a crepant resolution $f : Y \to X$. (Recall that a birational map of Gorenstein normal varieties is *crepant* if $f^* K_X = K_Y$. The unique person who could possibly come up with such jocular terms – standing for "zero discrepancies" – is Miles Reid.) Thus, string theorists posited that period integrals on the moduli space of Y equal Gromov–Witten numbers of Q, and vice versa.

One direction – relating period integrals of Y with Gromov–Witten invariants of Q, was proven, through elaborate direct computations, independently by Givental [23] and Lian-Liu-Yau [32]. Unfortunately this endeavor is marred by a disgusting and rather unnecessary controversy, which I will avoid altogether. One important point here is that the period integrals on the moduli space complex structures on Y coincide almost trivially with those of X, as any deformation of X persists with the same quotient singularities and the crepant resolution deforms along. So there is not a great issue of passing from X, or $\mathcal{X}$, to Y.

The other side is different: do we want to compare period integrals for Q with Gromov–Witten invariants of $\mathcal{X}$, or of Y? The first question that comes up is, does there exist a crepant resolution? The answer for threefolds is yes, one way to show it is using Nakajima's G-Hilbert scheme – and my favorite approach is that of Bridgeland, King and Reid [9]. In higher dimensions crepant resolutions may not exist and one needs to work directly on $\mathcal{X}$. It was Tseng in his thesis [38] who addressed the mirror prediction computations directly on the orbifold.

But the question remains, should we work with $\mathcal{X}$ or Y, and which Y for that matter? String theorists claim that it doesn't matter – Mirror symmetry works for both. We get the following *crepant resolution conjecture*:

Conjecture 6.6.1 (Ruan). *The Gromov–Witten theories of $\mathcal{X}$ and Y are equivalent.*

Moreover, it is conjectured that an explicit procedure of a particular type allows passing from X to Y and back.

This was Ruan's original impetus for developing orbifold cohomology!

There is a growing body of work on this exciting subject. See Li–Qin–Wang [31] and Bryan–Graber–Pandharipande [10].

A The Legend of String Cohomology: Two Letters of Maxim Kontsevich to Lev Borisov

A.1 The Legend of String Cohomology

On December 7, 1995 Kontsevich delivered a history-making lecture at Orsay, titled *String Cohomology*. "String cohomology" is the name Kontsevich chose to give what we know now, after Chen-Ruan, as *orbifold cohomology*, and Kontsevich's lecture notes described the orbifold and quantum cohomology of a global quotient orbifold. Twisted sectors, the age grading, and a version of orbifold stable maps for global quotients are all there.

Kontsevich never did publish his work on string cohomology. I met him in 1999 to discuss my work with Vistoli on twisted stable maps, and he told me a few hints about the Gromov–Witten theory of orbifolds (including the fact that the cohomology of the inertia stack is the model for quantum cohomology of X) which I could not appreciate at the time. He was also aware that Chen and Ruan were pursuing the subject so he had no intention to publish. We do, however, have written evidence in the form of two electronic mail letters he sent Lev Borisov in July 1996. These letters are reproduced below – verbatim with the exception of typographics – with Kontsevich and Borisov's permission.

What's more – Kontsevich never lectured about string cohomology after all! When I said he delivered a history-making lecture at Orsay, titled *String Cohomology*, I did not lie. The way François Loeser tells the story, this seems like an instance of the Heisenberg Uncertainty Principle: Loeser heard at the time Kontsevich had discovered a complex analogue of p-adic integration, and called Kontsevich to ask about it. In response Kontsevich told him to attend his December 7 lecture at Orsay on String Cohomology. Thus the lecture Kontsevich did deliver was indeed a history-making lecture on motivic integration.

For years people speculated what is this "String Cohomology", which they supposed Kontsevich aimed to develop out of Motivic Integration ....

Needless to say, Kontsevich never did publish his work on motivic integration either. A four-page set of notes is available in [29]. Of course that theory is by now fully developed.

A.2 The Archaeological Letters

Following are the two electronic mail letters of Kontsevich to Borisov from July 1996. I have trimmed the mailer headings, and, following Kontsevich's request, corrected typographical errors. I also added typesetting commands. Otherwise the text is Kontsevich's text verbatim.

The theory sketched here is the Gromov–Witten theory of a global quotient orbifold, a theory treated in great detail in [25].

Letter of 23 July 1996

Date: Tue, 23 Jul 96 12:39:42 +0200
From: maxim@ihes.fr (Maxim Kontsevich)
To: lborisov@msri.org
Subject: Re: stable maps to orbifolds

Dear Lev,

I didn't write yet anywhere the definition of stable maps to orbifolds, although I am planning to do it.

Here is it:

1. For an orbifold $X = Y/G$ (G is a finite group) we define a new orbifold X_1 as the quotient of

$$Y_1 := \{(x, g) | x \in X, g \in G, gx = x\}$$

   by the action of G:

$$f(x, g) := (fx, fgf^{-1}).$$

   Y_1 is a manifold consisting of parts of different dimensions, including Y itself (for $g = 1$).

2. String cohomology of X, say $HS(X)$ are defined as usual rational cohomology of X_1. I consider it for a moment only as a $\mathbb{Z}/2$-graded vector space (super space).

   One can also easily define HS for orbifolds which are not global quotients of smooth manifolds.

3. Let Y be an almost complex manifold. Denote by $\mathrm{Curves}(Y)$ the stack of triples (C, S, ϕ) where
   - C is a nonempty compact complex curve with may be double points.
   - C is not necessarily connected.
   - S is a finite subset of the smooth part of C.
   - ϕ is a holomorphic map from C to Y.

   Stability condition is the absence of infinitesimal automorphisms.

   The stack $\mathrm{Curves}(Y)$ consists of infinitely many components which are products of symmetric powers of usual moduli stacks of stable maps.

4. We define Pre-Curves(X) for $X = Y/G$ as the stack quotient of Curves(Y) by the obvious action of G.[2] Precisely it means that we consider quadruples (C, S, ϕ, A) where first three terms are as above and A is the action of G on (C, S, ϕ). In other words, we consider curves with G action and marked points and equivariant maps to X. We define Curves(X) as the subset of Pre-Curves(X) consisting of things where G acts FREELY on C minus {singular points and smooth points}. This is the essential part of the definition.[3]

   Lemma: Curves(X) is open and closed substack in Pre-Curves(X).

   As usual we can define the virtual fundamental class of each connected component of Curves(X).

5. For a curve from Curves(X) we can say when it is "connected": when the quotient curve C/G is connected. For "connected" curves we define their genus as genus of C/G, and the set of marked points as S/G. We define numbered stable map to X as a "connected" curve with numbered "marked points".

The evaluation map from the stable curve with numbered marked points takes value in the auxiliary orbifold X_1 (see 1, 2).

Thus we have a lot of symmetric tensors in $HS(X)$, and they all satisfy axioms from my paper with Manin. The interesting thing is the grading on HS:

It comes from the dimensions of virtual fundamental cycles. After working out the corresponding formula I get that one should define a new $\mathbb{Q}$-grading on $HS(X)$.

If Z is a connected component of X_1 we will define the rational number

$$age(Z) = \frac{1}{2\pi i} \sum (\log \text{ eigenvalues of the action of } g \text{ on } Tx),$$

where (x, g) is any point from Z.

The new $\mathbb{Q}$-degree of the component $H^k(Z)$ of $HS(X)$ I define as

$$k \;\; + \;\; 2 * age.$$

Notice that the $\mathbb{Z}/2$ grading (superstructure) is the old one, and has now nothing to do with the $\mathbb{Q}$-grading. Poincaré duality on HS comes from dualities on each $H(Z)$.

That's all, and I am a bit tired of typing by now.

Please write me if something is not clear.

There are some useful examples:

---

[2] This is not the standard notion of stack quotient. The next sentence explains what Kontsevich meant – D.A.

[3] Kontsevich omitted here the condition that G should not switch the branches at a node and have balanced action. It is automatic on things that can be smoothed – D.A.

$Y = $ 2dim torus, $G = Z/2$ with antipodal action.
Then $rk(HS) = 6$.
Another case: $Y = $ (complex surface $S)^n$, $G = $ symmetric group S_n.
$HS(Y) = H$(Hilbert scheme of S
　　　　　　　　resolving singularities in the symmetric power of S).[4]
All the best,
Maxim Kontsevich

Letter of 31 July 1996

Date: Wed, 31 Jul 96 11:21:35 +0200
From: maxim@ihes.fr (Maxim Kontsevich)
To: lborisov@msri.org
Subject: Re: stable maps to orbifolds

Dear Lev,
First of all, there was a misprint in the definition of X_1 as you noticed.[5]
I forgot in my letter to give the definition of the evaluation map. Namely,
if S_1 is an orbit of G acting on the curve, we chose first a point p from S_1.
The stabilizer A of p is a cyclic group because it acts freely on a punctured
neighborhood of p. Also, this cyclic group has a canonical generator g which
rotates the tangent space to p to the minimal angle in the anti-clockwise
direction. I associate with the orbit S_1 the point $(\phi(p), g)$ modulo G in X_1.
This point is independent of the choice of p in S_1.
Best,
Maxim Kontsevich
P.S. Please call me Maxim, not Professor

References

1. D. Abramovich, A. Corti and A. Vistoli, *Twisted bundles and admissible covers.*
   Special issue in honor of Steven L. Kleiman. Comm. Algebra 31 (2003), no. 8,
   3547–3618.
2. D. Abramovich, T. Graber and A. Vistoli, *Algebraic orbifold quantum products.*
   Orbifolds in mathematics and physics (Madison, WI, 2001), 1–24, Contemp.
   Math., 310, Amer. Math. Soc., Providence, RI, 2002.
3. –, *Gromov–Witten theory of Deligne–Mumford stacks*, preprint. arXiv:math/
   0603151, Amer. J. Math., to appear.

---

[4] Isomorphisms as groups, the relationship of ring structures is delicate – D.A.

[5] Borisov pointed out that an evaluation map to the non-rigidified inertia stack
$\mathcal{I}(\mathcal{X})$ (here X_1), and even just to $\mathcal{X}$, may not exist. Kontsevich defines it in what
follows on the level of coarse moduli spaces, which works fine – D.A.

4. D. Abramovich and A. Vistoli, *Complete moduli for fibered surfaces.* Recent progress in intersection theory (Bologna, 1997), 1–31, Trends Math., Birkhäuser Boston, Boston, MA, 2000.

5. –, *Compactifying the space of stable maps.* J. Amer. Math. Soc. 15 (2002), no. 1, 27–75.

6. K. Behrend, *Gromov-Witten invariants in algebraic geometry.* (English. English summary) Invent. Math. 127 (1997), no. 3, 601–617.

7. K. Behrend, and B. Fantechi, *The intrinsic normal cone.* Invent. Math. 128 (1997), no. 1, 45–88.

8. L. Borisov, L. Chen and G. Smith, *The orbifold Chow ring of toric Deligne-Mumford stacks.* J. Amer. Math. Soc. 18 (2005), no. 1, 193–215

9. T. Bridgeland, A. King, M. Reid, *Mukai implies McKay: The McKay correspondence as an equivalence of derived categories.* J. Amer. Math. Soc. 14 (2001), no. 3, 535–554

10. J. Bryan, T. Graber, and R. Pandharipande, *The orbifold quantum cohomology of C^2/Z_3 and Hurwitz-Hodge integrals*, J. Algebraic Geom. 17 (2008), no. 1, 1–28.

11. C. Cadman *Using stacks to impose tangency conditions on curves*, Amer. J. Math. 129 (2007), no. 2, 405–427.

12. C. Cadman *On the enumeration of rational plane curves with tangency conditions*, preprint math.AG/0509671

13. P. Candelas, X. de la Ossa, P. Green, and L. Parkes, *A pair of Calabi-Yau manifolds as an exactly soluble superconformal theory.* Nuclear Phys. B 359 (1991), no. 1, 21–74.

14. Chen B. and Hu S., *A deRham model for Chen-Ruan cohomology ring of abelian orbifolds*, Math. Ann. 336 (2006), no. 1, 51–71.

15. Chen W. and Ruan Y., *Orbifold Gromov-Witten theory.* Orbifolds in mathematics and physics (Madison, WI, 2001), 25–85, Contemp. Math., 310, Amer. Math. Soc., Providence, RI, 2002.

16. L. Dixon, J.A. Harvey, C. Vafa, E. Witten, *Strings on orbifolds.* Nuclear Phys. B 261 (1985), no. 4, 678–686.

17. L. Dixon, J.A. Harvey, C. Vafa, E. Witten, *Strings on orbifolds II.* Nuclear Phys. B 274 (1986), no. 2, 285–314.

18. D. Edidin, *Notes on the construction of the moduli space of curves*, Recent progress in intersection theory (Bologna, 1997), 85–113, Trends Math., Birkhäuser Boston, Boston, MA, 2000.

19. T. Ekedahl, *Boundary behaviour of Hurwitz schemes.* The moduli space of curves (Texel Island, 1994), 173–198, Progr. Math., 129, Birkhäuser Boston, Boston, MA, 1995.

20. B. Fantechi and L. Göttsche, *Orbifold cohomology for global quotients.* Duke Math. J. 117 (2003), no. 2, 197–227.

21. W. Fulton and R. Pandharipande, *Notes on stable maps and quantum cohomology*, in *Algebraic geometry—Santa Cruz 1995*, 45–96, Proc. Sympos. Pure Math., Part 2, Amer. Math. Soc., Providence, RI, 1997.

22. H. Gillet, *Intersection theory on algebraic stacks and Q-varieties*, J. Pure Appl. Algebra 34, 193-240 (1984).

23. A. Givental, *Equivariant Gromov-Witten invariants.* Internat. Math. Res. Notices 1996, no. 13, 613–663.

24. R. Goldin, T. Holm and A. Knutson, *Orbifold cohomology of torus quotients*, Duke Math. J. 139 (2007), no. 1, 89–139.

25. T. Jarvis, R. Kaufmann, and T. Kimura, *Pointed admissible G-covers and G-equivariant cohomological field theories.* Compos. Math. 141 (2005), no. 4, 926–978.

26. T. Jarvis, R. Kaufmann, and T. Kimura, *Stringy K-theory and the Chern character,* Invent. Math. 168 (2007), no. 1, 23–81.

27. S. Keel and S. Mori, *Quotients by groupoids,* Ann. of Math. (2) **145** (1997), no. 1, 193–213.

28. M. Kontsevich, *Enumeration of rational curves via torus actions.* The moduli space of curves (Texel Island, 1994), 335–368, Progr. Math., 129, Birkhäuser Boston, Boston, MA, 1995.

29. M. Kontsevich, *Grothendieck ring of motives and related rings,* notes for the lecture "String Cohomology" at Orsay, December 7, 1995. http://www.mabli.org/jet-preprints/Kontsevich-MotIntNotes.pdf

30. M. Kontsevich and Yu. Manin, *Gromov-Witten classes, quantum cohomology, and enumerative geometry.* Comm. Math. Phys. 164 (1994), no. 3, 525–562.

31. Li W.-P., Qin Z., and Wang W., *The cohomology rings of Hilbert schemes via Jack polynomials,* Algebraic structures and moduli spaces, 249–258, CRM Proc. Lecture Notes, 38, Amer. Math. Soc., Providence, RI, 2004.

32. Lian B., Liu K., Yau S.-T., *Mirror principle I* Asian J. Math. 1 (1997), no. 4, 729–763.

33. S. Mochizuki, *Extending families of curves over log regular schemes.* J. Reine Angew. Math. 511 (1999), 43–71.

34. M. Olsson, *Sheaves on Artin stacks,* J. Reine Angew. Math. 603 (2007), 55–112.

35. –, *Hom–stacks and restriction of scalars,* Duke Math. J. 134 (2006), no. 1, 139–164.

36. –, *On (log) twisted curves,* Compos. Math. 143 (2007), no. 2, 476–494.

37. M. Romagny, *Group actions on stacks and applications.* Michigan Math. J. 53 (2005), no. 1, 209–236.

38. Tseng H.-H., *Orbifold Quantum Riemann–Roch, Lefschetz and Serre,* preprint math.AG/0506111

39. B. Uribe, *Orbifold cohomology of the symmetric product.* Comm. Anal. Geom. 13 (2005), no. 1, 113–128.

40. A. Vistoli, *Intersection theory on algebraic stacks and on their moduli spaces.* Invent. Math. 97 (1989), no. 3, 613-670.

41. E. Zaslow, *Topological orbifold models and quantum cohomology rings.* Comm. Math. Phys. 156 (1993), no. 2, 301–331.

Lectures on the Topological Vertex

M. Mariño[*]

Department of Physics, Theory Division, CERN, University of Geneva
1211 Geneva, Switzerland
marcos@mail.cern.ch
Marcos.Marino.Beiras@cern.ch

1 Introduction and Overview

The theory of Gromov–Witten invariants was largely motivated by the study of string theory on Calabi–Yau manifolds, and has now developed into one of the most dynamic fields of algebraic geometry. During the last years there has been enormous progress in the development of the theory and of its computational techniques. Roughly speaking, and restricting ourselves to Calabi–Yau threefolds, we have the following mathematical approaches to the computation of Gromov–Witten invariants:

1. *Localization.* This was first proposed by Kontsevich, and requires torus actions in the Calabi–Yau in order to work. Localization provides a priori a complete solution of the theory on toric (hence non-compact) Calabi–Yau manifolds, and reduces the computation of Gromov–Witten invariants to the calculation of Hodge integrals in Deligne–Mumford moduli space. Localization techniques make also possible to solve the theory at genus zero on a wide class of compact manifolds, see for example Cox and Katz (1999) for a review.
2. *Deformation and topological approach.* This has been developed more recently and relies on *relative* Gromov–Witten invariants. It provides a cut-and-paste approach to the calculation of the invariants and seems to be the most powerful approach to higher genus Gromov–Witten invariants in the compact case.
3. *D-brane moduli spaces.* Gromov–Witten invariants can be reformulated in terms of the so-called Gopakumar–Vafa invariants (see Hori et al. (2003) for a summary of these). Heuristic techniques to compute them in terms of Euler characteristics of moduli space of embedded surfaces, and one can recover to a large extent the original information of Gromov–Witten theory. The equivalence between these two invariants remains however

---

[*] Also at Departamento de Matemática, IST, Lisboa, Portugal

K. Behrend, M. Manetti (eds.), *Enumerative Invariants in Algebraic Geometry and String Theory.* Lecture Notes in Mathematics 1947,
© Springer-Verlag Berlin Heidelberg 2008

conjectural, and a general, rigorous definition of the Gopakumar–Vafa invariants in terms of appropriate moduli spaces is still not known. There is another set of invariants, the so-called *Donaldson–Thomas invariants*, that are also related to D-brane moduli spaces, which can be rigorously defined and have been conjectured to be equivalent to Gromov–Witten invariants by Maulik, Nekrasov, Okounkov and Pandharipande (2003).

Gromov–Witten invariants are closely related to string theory. It turns out that type IIA theory on a Calabi–Yau manifold X leads to a four-dimensional supersymmetric theory whose Lagrangian contains moduli-dependent couplings $F_g(t)$, where t denotes the Kähler moduli of the Calabi–Yau. When these couplings are expanded in the large radius limit, they are of the form

$$(1) \qquad F_g(t) = \sum_{\beta \in H_2(X)} N_{g,\beta}\, e^{-\beta \cdot t},$$

where $N_{g,\beta}$ are the Gromov–Witten invariants for the class β at genus g (see Sect. 3 below for details on this). It turns out that there is a simplified version of string theory, called topological string theory, which captures precisely the information contained in these couplings. Topological string theory comes in two versions, called the A and the B model (see Hori et al. (2003) and Mariño (2005) for a review). Type A topological string theory is related to Gromov–Witten theory, and its free energy at genus g is precisely given by (1). Type B topological string theory is related to the deformation theory of complex structures of the Calabi–Yau manifold. In the last years, various dualities of string theory have led to powerful techniques to compute these couplings, hence Gromov–Witten invariants:

1. *Mirror symmetry.* Mirror symmetry relates type A theory on a Calabi–Yau manifold X to type B theory on the mirror manifold $\widetilde{X}$. When the mirror of the Calabi–Yau X is known, this leads to a complete solution at genus zero in terms of variation of the complex structures of $\widetilde{X}$. For genus $g \geq 1$, mirror symmetry can be combined with the holomorphic anomaly equations of Bershadsky et al. (1994) to obtain $F_g(t)$. However, this does not provide the full solution to the model due to the so-called holomorphic ambiguity. On the other hand, mirror symmetry and the holomorphic anomaly equation are very general and work for both compact and non-compact Calabi–Yau manifolds.

2. *Large N dualities.* Large N dualities lead to a computation of the $F_g(t)$ couplings in terms of correlation functions and partition functions in Chern–Simons theory. Although this was formulated originally only for the resolved conifold, one ends up with a general theory – the theory of the topological vertex, introduced in Aganagic et al. (2005) – which leads to a complete solution on toric Calabi–Yau manifolds. The theory of the topological vertex is closely related to localization and to Hodge integrals, and it can be formulated in a rigorous mathematical way (see Li et al. 2004).

3. *Heterotic duality.* When the Calabi–Yau manifold has the structure of a K3 fibration, type IIA theory often has a heterotic dual, and the evaluation of $F_g(t)$ restricted to the K3 fiber can be reduced to a one-loop integral in heterotic string theory. This leads to explicit, conjectural formulae for Gromov–Witten invariants in terms of modular forms.

In this lectures, I will summarize the approach to Gromov–Witten invariants on toric Calabi–Yau threefolds based on large N dualities. Since the large N duality/topological vertex approach computes Gromov–Witten invariants in terms of Chern–Simons knot and link invariants, Sect. 2 is devoted to a review of these. Section 3 reviews topological strings and Gromov–Witten invariants, and gives some information about the open string case. Section 4 introduces the class of geometries we will deal with, namely toric (noncompact) Calabi–Yau manifolds, and we present a useful graphical way to represent these manifolds which constitutes the geometric core of the theory of the topological vertex. Finally, in Sect. 5, we define the vertex and present some explicit formulae for it and some simple applications. A brief Appendix contains useful information about symmetric polynomials.

It has not been possible to present all the relevant background and physical derivations in this set of lectures. However, these topics have been extensively reviewed for example in the book Mariño (2005), to which we refer for further information and/or references.

2 Chern–Simons Theory

2.1 Basic Ingredients

In a groundbreaking paper, Witten (1989) showed that Chern–Simons gauge theory, which is a quantum field theory in three dimensions, provides a physical description of a wide class of invariants of three-manifolds and of knots and links in three-manifolds.[1] The Chern–Simons action with gauge group G on a generic three-manifold M is defined by

$$(2) \qquad S = \frac{k}{4\pi} \int_M \mathrm{Tr}\left(A \wedge dA + \frac{2}{3} A \wedge A \wedge A\right).$$

Here, k is the coupling constant, and A is a G-gauge connection on the trivial bundle over M. In the following, we will mostly consider Chern–Simons theory with gauge group $G = U(N)$.

Chern–Simons theory is an example of a *topological field theory*. The reason is that the Chern–Simons theory action does not involve the metric of M in order to be defined, and the partition function

$$(3) \qquad Z(M) = \int [\mathcal{D}A] e^{iS}$$

---

[1] This was also conjectured by Schwarz (1987).

should define a topological invariant of the manifold M. The fact, however, that the classical Lagrangian is metric independent is not, in general, sufficient to guarantee that the quantum theory will preserve this invariance, since there could be anomalies in the quantization process that spoil the classical symmetry. A detailed analysis due to Witten (1989) shows that, in the case of Chern–Simons theory, topological invariance is preserved quantum mechanically, but with an extra subtlety: the invariant depends not only on the three-manifold but also on a choice of framing, i.e. a choice of trivialization of the bundle $TM \oplus TM$. The choice of framing changes the value of the partition function in a very precise way: if the framing is changed by n units, the partition function $Z(M)$ changes as follows:

(4)
$$Z(M) \to \exp\left[\frac{\pi i n c}{12}\right] Z(M),$$

where

(5)
$$c = \frac{kd}{k+y}.$$

In this equation, d and y are, respectively, the dimension and the dual Coxeter number of the group G (for $G = U(N)$, $y = N$). As explained by Atiyah (1990), for every three-manifold there is in fact a canonical choice of framing, and the different choices are labelled by an integer $s \in \mathbf{Z}$ in such a way that $s = 0$ corresponds to the canonical framing. In the following, unless otherwise stated, all the results for the partition functions of Chern–Simons theory will be presented in the canonical framing.

Besides providing invariants of three-manifolds, Chern–Simons theory also provides invariants of knots and links inside three-manifolds (for a survey of modern knot theory, see Lickorish 1998, and Prasolov and Sossinsky 1997). Some examples of knots and links are depicted in Fig. 1. Given an oriented knot $\mathcal{K}$ in $\mathbf{S}^3$, we can consider the trace of the holonomy of the gauge connection around $\mathcal{K}$ in a given irreducible representation R of $U(N)$. This gives the Wilson loop operator:

(6)
$$W_R^{\mathcal{K}}(A) = \mathrm{Tr}_R U_{\mathcal{K}},$$

where

(7)
$$U_{\mathcal{K}} = \mathrm{P} \exp \oint_{\mathcal{K}} A$$

is the holonomy around the knot. The operator in equation (6) is a gauge-invariant operator whose definition does not involve the metric on the three-manifold, therefore it is an observable of Chern–Simons theory regarded as a topological field theory. The irreducible representations of $U(N)$ will be labelled by highest weights or equivalently by the lengths of rows in a Young

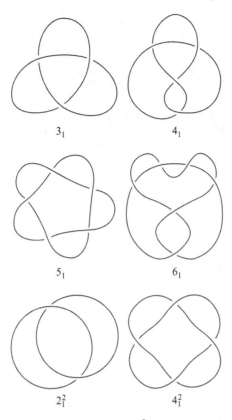

Fig. 1. Some knots and links. In the notation x_n^L, x indicates the number of crossings, L the number of components (when it is a link with $L > 1$) and n is a number used to enumerate knots and links in a given set characterized by x and L. The knot 3_1 is also known as the trefoil knot, while 4_1 is known as the figure-eight knot. The link 2_1^2 is called the Hopf link

tableau, l_i, where $l_1 \geq l_2 \geq \cdots$. If we now consider a link $\mathcal{L}$ with components $\mathcal{K}_\alpha$, $\alpha = 1, \cdots, L$, we can in principle compute the normalized correlation function,

$$(8) \quad W_{R_1 \cdots R_L}(\mathcal{L}) = \langle W_{R_1}^{\mathcal{K}_1} \cdots W_{R_L}^{\mathcal{K}_L} \rangle = \frac{1}{Z(M)} \int [\mathcal{D}A] \left(\prod_{\alpha=1}^{L} W_{R_\alpha}^{\mathcal{K}_\alpha} \right) e^{iS}.$$

The unnormalized correlation function will be denoted by $Z_{R_1 \cdots R_L}(\mathcal{L})$. The topological character of the action, and the fact that the Wilson loop operators can be defined without using any metric on the three-manifold, indicate that (8) is a topological invariant of the link $\mathcal{L}$. Similarly to what happens with the partition function, in order to define the invariant of the link we need some extra information due to quantum ambiguities in the correlation function (8).

For further use we notice that, given two linked oriented knots $\mathcal{K}_1$, $\mathcal{K}_2$, one can define an elementary topological invariant, the *linking number*, by

(9) $$\mathrm{lk}(\mathcal{K}_1, \mathcal{K}_2) = \frac{1}{2} \sum_p \epsilon(p),$$

where the sum is over all crossing points, and $\epsilon(p) = \pm 1$ is a sign associated to the crossings as indicated in Fig. 2. The linking number of a link $\mathcal{L}$ with components $\mathcal{K}_\alpha$, $\alpha = 1, \cdots, L$, is defined by

(10) $$\mathrm{lk}(\mathcal{L}) = \sum_{\alpha < \beta} \mathrm{lk}(\mathcal{K}_\alpha, \mathcal{K}_\beta).$$

For example, once an orientation is chosen for the two components of the Hopf

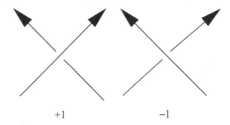

$$+1 \qquad\qquad -1$$

Fig. 2. When computing the linking number of two knots, the crossings are assigned a sign ± 1 as indicated in the figure

link 2_1^2 shown in Fig. 1, one finds two inequivalent oriented links with linking numbers ± 1.

Some of the correlation functions of Wilson loops in Chern–Simons theory turn out to be closely related to important polynomial invariants of knots and links. For example, one of the most important polynomial invariants of a link $\mathcal{L}$ is the HOMFLY polynomial $P_{\mathcal{L}}(q, \lambda)$, which depends on two variables q and λ and was introduced by Freyd et al. (1985). This polynomial turns out to be related to the correlation function (8) when the gauge group is $U(N)$ and all the components are in the fundamental representation $R_\alpha = \square$. More precisely, we have

(11) $$W_{\square \cdots \square}(\mathcal{L}) = \lambda^{\mathrm{lk}(\mathcal{L})} \left(\frac{\lambda^{\frac{1}{2}} - \lambda^{-\frac{1}{2}}}{q^{\frac{1}{2}} - q^{-\frac{1}{2}}} \right) P_{\mathcal{L}}(q, \lambda)$$

where $\mathrm{lk}(\mathcal{L})$ is the linking number of $\mathcal{L}$, and the variables q and λ are related to the Chern–Simons variables as

(12) $$q = e^x, \quad x = \frac{2\pi i}{k + N}, \quad \lambda = q^N.$$

When $N = 2$ the HOMFLY polynomial reduces to a one-variable polynomial, the Jones polynomial. When the gauge group of Chern–Simons theory is $SO(N)$, $W_{\square\cdots\square}(\mathcal{L})$ is closely related to the Kauffman polynomial. For the mathematical definition and properties of these polynomials, see, for example, Lickorish (1998).

2.2 Perturbative Approach

The partition function and correlation functions of Wilson loops in Chern–Simons theory can be computed in a variety of ways. We will here present the basic results of Chern–Simons perturbation theory for the partition function. Since our main interest will be the non-perturbative results of Witten (1989), we will be rather sketchy. For more information on Chern–Simons perturbation theory, we refer the reader to Dijkgraaf (1995) and Labastida (1999) for a physical point of view, and Bar-Natan (1995) and Ohtsuki (2002), for a mathematical perspective.

In the computation of the partition function in perturbation theory, we have first to find the classical solutions of the Chern–Simons equations of motion. If we write $A = \sum_a A^a T_a$, where T_a is a basis of the Lie algebra, we find

$$\frac{\delta S}{\delta A_\mu^a} = \frac{k}{4\pi} \epsilon^{\mu\nu\rho} F_{\nu\rho}^a,$$

therefore the classical solutions are just flat connections on M. Flat connections are in one-to-one correspondence with group homomorphisms

$$(13) \qquad\qquad \pi_1(M) \to G.$$

For example, if $M = \mathbf{S}^3/\mathbf{Z}_p$ is the lens space $L(p, 1)$, one has $\pi_1(L(p, 1)) = \mathbf{Z}_p$, and flat connections are labelled by homomorphisms $\mathbf{Z}_p \to G$. Let us assume that the flat connections on M are a discrete set of points (this happens, for example, if M is a rational homology sphere, since in that case $\pi_1(M)$ is a finite group). In that situation, one expresses $Z(M)$ as a sum of terms associated to stationary points:

$$(14) \qquad\qquad Z(M) = \sum_c Z^{(c)}(M),$$

where c labels the different flat connections $A^{(c)}$ on M. Each of the $Z^{(c)}(M)$ will be an asympotic series in $1/k$ of the form

$$(15) \qquad\qquad Z^{(c)}(M) = Z_{1-\text{loop}}^{(c)}(M) \exp\left\{ \sum_{\ell=1}^{\infty} S_\ell^{(c)} x^\ell \right\}.$$

In this equation, x is the effective expansion parameter:

$$(16) \qquad\qquad x = \frac{2\pi i}{k+y},$$

which takes into account a quantum shift $k \to k+y$ due to finite renormalization effects. The one-loop correction $Z_{1-\text{loop}}^{(c)}(M)$ was first analyzed by Witten (1989), and has been studied in great detail since then (Freed and Gompf 1991; Jeffrey 1992; Rozansky 1995). It has the form

$$(17) \qquad Z_{1-\text{loop}}^{(c)}(M) = \frac{(2\pi x)^{\frac{1}{2}(\dim H_c^0 - \dim H_c^1)}}{\text{vol}(H_c)} e^{-\frac{1}{x}S_{\text{CS}}(A^{(c)}) - \frac{i\pi}{4}\varphi} \sqrt{|\tau_R^{(c)}|},$$

where $H_c^{0,1}$ are the cohomology groups with values in the Lie algebra of G associated to the flat connection $A^{(c)}$, $\tau_R^{(c)}$ is the Reidemeister–Ray–Singer torsion of $A^{(c)}$, H_c is the isotropy group of $A^{(c)}$, and φ is a certain phase. Notice that, for the trivial flat connection $A^{(c)} = 0$, $H_c = G$.

Let us focus on the terms in (15) corresponding to the trivial connection, which will be denoted by S_ℓ. Diagramatically, the free energy is computed by connected bubble diagrams made out of trivalent vertices (since the interaction in the Chern–Simons action is cubic). We will refer to these diagrams as *connected trivalent graphs*. S_ℓ is the contribution of connected trivalent graphs with 2ℓ vertices and $\ell + 1$ loops. For each of these graphs we have to compute a group factor and a Feynman integral. However, not all these graphs are independent, since the underlying Lie algebra structure imposes the Jacobi identity:

$$(18) \qquad \sum_e \left(f_{abe}f_{edc} + f_{dae}f_{ebc} + f_{ace}f_{edb} \right) = 0.$$

This leads to the graph relation known as the IHX relation. Also, antisymmetry of f_{abc} leads to the so-called AS relation (see, for example, Bar-Natan 1995; Dijkgraaf 1995; Ohtsuki 2002). The existence of these relations suggests to define an equivalence relation in the space of connected trivalent graphs by quotienting by the IHX and the AS relations, and this gives the so-called *graph homology*. The space of homology classes of connected diagrams will be denoted by $\mathcal{A}(\emptyset)^{\text{conn}}$. This space is graded by half the number of vertices ℓ, and this number gives the degree of the graph. The space of homology classes of graphs at degree ℓ is then denoted by $\mathcal{A}(\emptyset)_\ell^{\text{conn}}$. For every ℓ, this is a finite-dimensional vector space of dimension $d(\ell)$. The dimensions of these spaces are explicitly known for low degrees, see, for example, Bar-Natan (1995), and we have listed some of them in Table 1. Given any group G, we have a map

$$(19) \qquad r_G : \mathcal{A}(\emptyset)^{\text{conn}} \longrightarrow \mathbf{R}$$

that associates to every graph Γ its group theory factor $r_G(\Gamma)$. This map is of course well defined, since different graphs in the same homology class $\mathcal{A}(\emptyset)^{\text{conn}}$ lead by definition to the same group factor. This map is an example of a *weight system* for $\mathcal{A}(\emptyset)^{\text{conn}}$. Every gauge group gives a weight system for $\mathcal{A}(\emptyset)^{\text{conn}}$, but one may, in principle, find weight systems not associated to gauge groups, although so far the only known example is the one constructed by Rozansky and Witten (1997), which instead uses hyperKähler manifolds. We can now state very precisely what is the structure of the S_ℓ appearing in (15): since the Feynman diagrams can be grouped into homology classes, we have

Table 1. Dimensions $d(\ell)$ of $\mathcal{A}(\emptyset)_\ell^{\text{conn}}$ up to $\ell = 10$

ℓ	1	2	3	4	5	6	7	8	9	10
$d(\ell)$	1	1	1	2	2	3	4	5	6	8

$$(20) \qquad S_\ell = \sum_{\Gamma \in \mathcal{A}(\emptyset)_\ell^{\text{conn}}} r_G(\Gamma) I_\Gamma(M).$$

The factors $I_\Gamma(M)$ appearing in (20) are certain sums of integrals of propagators over M. It was shown by Axelrod and Singer (1992) that these are differentiable invariants of the three-manifold M, and since the dependence on the gauge group has been factored out, they only capture topological information of M, in contrast to $Z(M)$, which also depends on the choice of the gauge group. These are the *universal perturbative invariants* defined by Chern–Simons theory. Notice that, at every order ℓ in perturbation theory, there are $d(\ell)$ independent perturbative invariants. Of course, these invariants inherit from $\mathcal{A}(\emptyset)_\ell^{\text{conn}}$ the structure of a finite-dimensional vector space, and in particular one can choose a basis of trivalent graphs. A possible choice for $\ell \leq 5$ is the following (Sawon 2004):

$$
\begin{aligned}
\ell &= 1: \quad . \\
\ell &= 2: \quad . \\
\ell &= 3: \quad . \\
\ell &= 4: \quad . \quad . \\
\ell &= 5: \quad . \quad . .
\end{aligned}
$$

(21)

We will denote the graphs with k circles joined by lines by θ_k. Therefore, the graph corresponding to $\ell = 1$ will be denoted by θ, the graph corresponding to $\ell = 2$ will be denoted θ_2, and so on.

Notice that Chern–Simons theory detects the graph homology through the weight system associated to Lie algebras, so in principle it could happen that there is an element of graph homology that is not detected by these weight systems. There is, however, a very elegant mathematical definition of the universal perturbative invariant of a three-manifold that works directly in the graph homology. This is called the LMO invariant (Le et al. 1998) and it is a formal linear combination of homology graphs with rational coefficients:

(22) $$\omega(M) = \sum_{\Gamma \in \mathcal{A}(\emptyset)^{\text{conn}}} I_\Gamma^{\text{LMO}}(M)\,\Gamma \in \mathcal{A}(\emptyset)^{\text{conn}}[\mathbf{Q}].$$

It is believed that the universal invariants extracted from Chern–Simons perturbation theory agree with the LMO invariant. More precisely, since the LMO invariant $\omega(M)$ is taken to be 0 for $\mathbf{S}^3$, we have:

(23) $$I_\Gamma^{\text{LMO}}(M) = I_\Gamma(M) - I_\Gamma(\mathbf{S}^3),$$

as long as the graph Γ is detected by Lie algebra weight systems. In that sense the LMO invariant is more refined than the universal perturbative invariants extracted from Chern–Simons theory; see Ohtsuki (2002) for a detailed introduction to the LMO invariant and its properties.

The computation of S_ℓ involves the evaluation of group factors of Feynman diagrams, which we have denoted by $r_G(\Gamma)$ above. Here, we give some details about how to evaluate these factors when $G = U(N)$, following the diagrammatic techniques of Cvitanovic (1976) and Bar-Natan (1995). A systematic discussion of these techniques can be found in Cvitanovic (2004).

Fig. 3. Graphic representation of the generator $(T_a)_{ij}$ of a Lie algebra

The basic idea to evaluate group factors is very similar to the double-line notation of 't Hooft (1974), and it amounts to expressing indices in the adjoint representation in terms of indices in the fundamental (and anti-fundamental) representation. The resulting diagrams are often called *fatgraphs*. In the case of $U(N)$, the adjoint representation is just the tensor product of the fundamental and the anti-fundamental representation. Let us first normalize the trace in the fundamental representation by setting

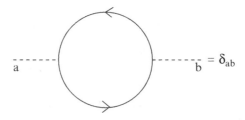

Fig. 4. Graphic representation of the normalization condition (24)

(24) $$\mathrm{Tr}\,(T_a\,T_b) = \delta_{ab}, \qquad a,b = 1, \cdots, N^2.$$

One can then see that

(25) $$\sum_a (T_a)_{ij}(T_a)_{kl} = \delta_{il}\delta_{kj}.$$

If we represent the generator $(T_a)_{ij}$ as in Fig. 3, the relation (25) can in turn be represented as Fig. 5. This is simply the statement that the adjoint representation of $U(N)$ is given by $V_N \otimes V_N^*$. Similarly, the normalization condition (24) is graphically represented as Fig. 4. The evaluation of group

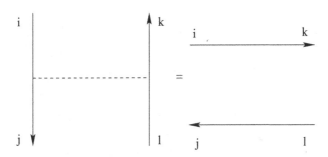

Fig. 5. Graphic representation of (25)

factors of Feynman diagrams involves, of course, the structure constants of the Lie algebra f_{abc}, associated to the cubic vertex. By tracing the defining relation of the structure constants we find

(26) $$f_{abc} = \mathrm{Tr}\,(T_a T_b T_c) - \mathrm{Tr}\,(T_b T_a T_c),$$

which we represent as Fig. 6. Putting this together with Fig. 5, we obtain the

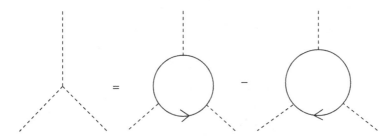

Fig. 6. Graphic representation of the relation (26) between structure constants and generators

graphical rule represented in Fig. 7. We can interpret this as a rule that tells us how to split a single-line Feynman diagram of the $U(N)$ theory into fatgraphs: given a Feynman diagram, we substitute each vertex by the double line vertex without twists, minus the double-line vertex with twists in all edges. If the diagram has 2ℓ vertices, we will generate 4^ℓ fatgraphs (some of them may be equal), with a $\pm$ sign, which can be interpreted as Riemann surfaces with holes. The group factor of a fatgraph with h holes is simply N^h.

Example. As an example of the above procedure, One can use the above rules to compute the group factor of the two-loop Feynman diagram

(27)

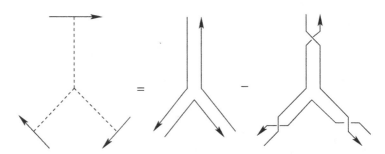

Fig. 7. Graphic rule to transform Feynman diagrams into double-line diagrams

By resolving the two vertices we obtain two different fatgraphs: the graph in Fig. 8 with weight 2, and the graph in Fig. 9 with weight -2. One then finds:

(28) $$r_{U(N)}(\theta) = 2N(N^2 - 1).$$

Similarly, the same procedure gives

(29) $$r_{U(N)}(\theta_2) = 4N^2(N^2 - 1).$$

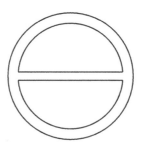

Fig. 8. A fatgraph obtained from the Feynman diagram (27)

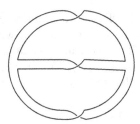

Fig. 9. Another fatgraph obtained from (27)

It is easy to see from the evaluation of group factors that the perturbative expansion of the free energy of Chern–Simons theory around the trivial connection can be written in the form

$$(30) \qquad F = \sum_{g=0}^{\infty}\sum_{h=1}^{\infty} F_{g,h}x^{2g-2+h}N^h.$$

In fact, this structure for the partition function holds for any quantum theory containing only fields in the adjoint representation ('t Hooft 1974). One can also reorganize the perturbative series (30) as

$$(31) \qquad F = \sum_{g=0}^{\infty} F_g(t)g_s^{2g-2},$$

where t is called the *'t Hooft coupling* of Chern–Simons theory and it is given by

$$(32) \qquad t = Nx,$$

and $F_g(t)$ is defined by summing over all holes keeping the genus g fixed:

$$(33) \qquad F_g(t) = \sum_{h=1}^{\infty} F_{g,h}t^h.$$

We will see later in this section how to compute the coefficients $F_{g,h}$ and the function $F_g(t)$ for Chern–Simons theory on $\mathbf{S}^3$.

2.3 Non-Perturbative Solution

As was shown by Witten (1989), Chern–Simons theory is exactly solvable by using non-perturbative methods and the relation to the Wess–Zumino–Witten (WZW) model. In order to present this solution, it is convenient to recall some basic facts about the canonical quantization of the model.

Let M be a three-manifold with boundary given by a Riemann surface Σ. We can insert a general operator $\mathcal{O}$ in M, which will, in general, be a product of Wilson loops along different knots and in arbitrary representations of the gauge group. We will consider the case in which the Wilson loops do not intersect the surface Σ. The path integral over the three-manifold with boundary M gives a wavefunction $\Psi_{M,\mathcal{O}}(\mathcal{A})$ that is a functional of the values of the field on Σ. Schematically, we have:

$$(34) \qquad \Psi_{M,\mathcal{O}}(\mathcal{A}) = \langle \mathcal{A} | \Psi_{M,\mathcal{O}} \rangle = \int_{A|_{\Sigma} = \mathcal{A}} \mathcal{D}A \, e^{iS} \, \mathcal{O}.$$

In fact, associated to the Riemann surface Σ we have a Hilbert space $\mathcal{H}(\Sigma)$, which can be obtained by doing canonical quantization of Chern–Simons theory on $\Sigma \times \mathbf{R}$. Before spelling out in detail the structure of these Hilbert spaces, let us make some general considerations about the computation of physical quantities.

In the context of canonical quantization, the partition function can be computed as follows. We first perform a Heegaard splitting of the three-manifold, i.e. we represent it as the connected sum of two three-manifolds M_1 and M_2 sharing a common boundary Σ, where Σ is a Riemann surface. If $f : \Sigma \to \Sigma$ is a homeomorphism, we will write $M = M_1 \cup_f M_2$, so that M is obtained by gluing M_1 to M_2 through their common boundary and using the homeomorphism f. This is represented in Fig. 10. We can then compute the full path integral (3) over M by computing first the path integral over M_1 to obtain a state $|\Psi_{M_1}\rangle$ in $\mathcal{H}(\Sigma)$. The boundary of M_2 is also Σ, but with opposite orientation, so its Hilbert space is the dual space $\mathcal{H}^*(\Sigma)$. The path integral over M_2 then produces a state $\langle \Psi_{M_2} | \in \mathcal{H}^*(\Sigma)$. The homeomorphism $f : \Sigma \to \Sigma$ will be represented by an operator acting on $\mathcal{H}(\Sigma)$,

$$(35) \qquad U_f : \mathcal{H}(\Sigma) \to \mathcal{H}(\Sigma).$$

and the partition function can be finally evaluated as

$$(36) \qquad Z(M) = \langle \Psi_{M_2} | U_f | \Psi_{M_1} \rangle.$$

Therefore, if we know explicitly what the wavefunctions and the operators associated to homeomorphisms are, we can compute the partition function. The result of the computation is, of course, independent of the particular Heegaard splitting of M.

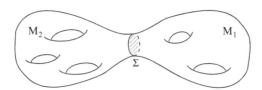

Fig. 10. Heegaard splitting of a three-manifold M into two three-manifolds M_1 and M_2 with a common boundary Σ

One of the most fundamental results of Witten (1989) is a precise description of $\mathcal{H}(\Sigma)$: it is the space of conformal blocks of a WZW model on Σ with gauge group G and level k (for an extensive review of the WZW model, see, for example, Di Francesco et al. 1997). In particular, $\mathcal{H}(\Sigma)$ has finite dimension. We will not review here the derivation of this fundamental result. Instead we will use the relevant information from the WZW model in order to solve Chern–Simons theory in some important cases.

The description of the space of conformal blocks on Riemann surfaces can be made very explicit when Σ is a sphere or a torus. For $\Sigma = \mathbf{S}^2$, the space of conformal blocks is one-dimensional, so $\mathcal{H}(\mathbf{S}^2)$ is spanned by a single element. For $\Sigma = \mathbf{T}^2$, the space of conformal blocks is in one-to-one correspondence with the integrable representations of the affine Lie algebra associated to G at level k. We will use the following notations: the fundamental weights of G will be denoted by λ_i, and the simple roots by α_i, $i = 1, \cdots, r$, where r denotes the rank of G. The weight and root lattices of G are denoted by Λ^w and Λ^r, respectively, and $|\Delta_+|$ denotes the number of positive roots. The fundamental chamber $\mathcal{F}_l$ is given by $\Lambda^w/l\Lambda^r$, modded out by the action of the Weyl group. For example, in $SU(N)$ a weight $p = \sum_{i=1}^{r} p_i \lambda_i$ is in $\mathcal{F}_l$ if

$$(37) \qquad \sum_{i=1}^{r} p_i < l, \quad \text{and} \quad p_i > 0, \, i = 1, \cdots, r.$$

We recall that a representation given by a highest weight Λ is integrable if $\rho + \Lambda$ is in the fundamental chamber $\mathcal{F}_l$, where $l = k+y$ (ρ denotes as usual the Weyl vector, given by the sum of the fundamental weights). In the following, the states in the Hilbert state of the torus $\mathcal{H}(\mathbf{T}^2)$ will be denoted by $|p\rangle = |\rho + \Lambda\rangle$ where $\rho + \Lambda \in \mathcal{F}_l$, as we have stated, is an integrable representation of the WZW model at level k. We will also denote these states by $|R\rangle$, where R is the representation associated to Λ. The state $|\rho\rangle$ will be denoted by $|0\rangle$. The states $|R\rangle$ can be chosen to be orthonormal (Witten 1989; Elitzur et al. 1989; Labastida and Ramallo 1989), so we have

$$(38) \qquad \langle R|R'\rangle = \delta_{RR'}.$$

There is a special class of homeomorphisms of $\mathbf{T}^2$ that have a simple expression as operators in $\mathcal{H}(\mathbf{T}^2)$; these are the $SL(2, \mathbf{Z})$ transformations. Recall that the group $SL(2, \mathbf{Z})$ consists of 2×2 matrices with integer entries and unit determinant. If $(1, 0)$ and $(0, 1)$ denote the two one-cycles of $\mathbf{T}^2$, we can specify the action of an $SL(2, \mathbf{Z})$ transformation on the torus by giving its action on this homology basis. The $SL(2, \mathbf{Z})$ group is generated by the transformations T and S, which are given by

$$(39) \qquad T = \begin{pmatrix} 1 & 1 \\ 0 & 1 \end{pmatrix}, \quad S = \begin{pmatrix} 0 & -1 \\ 1 & 0 \end{pmatrix}.$$

Notice that the S transformation exchanges the one-cycles of the torus. These transformations can be lifted to $\mathcal{H}(\mathbf{T}^2)$, and they have the following matrix elements in the basis of integrable representations:

$$T_{pp'} = \delta_{p,p'} e^{2\pi i(h_p - c/24)},$$

$$(40) \quad S_{pp'} = \frac{i^{|\Delta_+|}}{(k+y)^{r/2}} \left(\frac{\mathrm{Vol}\, \Lambda^w}{\mathrm{Vol}\, \Lambda^r} \right)^{\frac{1}{2}} \sum_{w \in W} \epsilon(w) \exp\left(-\frac{2\pi i}{k+y} p \cdot w(p') \right).$$

In the first equation, c is the central charge of the WZW model, and h_p is the conformal weight of the primary field associated to p:

$$(41) \qquad\qquad h_p = \frac{p^2 - \rho^2}{2(k+y)},$$

where we recall that p is of the form $\rho + \Lambda$. In the second equation, the sum over w is a sum over the elements of the Weyl group W, $\epsilon(w)$ is the signature of the element w, and $\mathrm{Vol}\, \Lambda^w (\mathrm{Vol}\, \Lambda^r)$ denote, respectively, the volume of the weight (root) lattice. We will often write $S_{RR'}$ for $S_{pp'}$, where $p = \rho + \Lambda$, $p' = \rho + \Lambda'$ and Λ, Λ' are the highest weights corresponding to the representations R, R'.

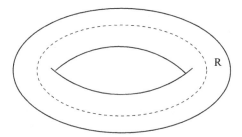

Fig. 11. Performing the path integral on a solid torus with a Wilson line in representation R gives the state $|R\rangle$ in $\mathcal{H}(\mathbf{T}^2)$

What is the description of the states $|R\rangle$ in $\mathcal{H}(\mathbf{T}^2)$ from the point of view of canonical quantization? Consider the solid torus $\mathcal{T} = D \times \mathbf{S}^1$, where D is a disc in $\mathbf{R}^2$. This is a three-manifold whose boundary is a $\mathbf{T}^2$, and it has a non-contractible cycle given by the $\mathbf{S}^1$. Let us now consider the Chern–Simons path integral on the solid torus, with the insertion of the operator $\mathcal{O}_R = \mathrm{Tr}_R U$ given by a Wilson loop in the representation R around the non-contractible cycle, as shown in Fig. 11. In this way, one obtains a state in $\mathcal{H}(\mathbf{T}^2)$, and one has

$$(42) \qquad\qquad |\Psi_{\mathcal{T},\mathcal{O}_R}\rangle = |R\rangle.$$

In particular, the path integral over the solid torus with no operator insertion gives $|0\rangle$, the 'vacuum' state.

These results allow us to compute the partition function of any three-manifold that admits a Heegaard splitting along a torus. Imagine, for example, that we take two solid tori and we glue them along their boundary with the identity map. Since a solid torus is a disc times a circle, $D \times \mathbf{S}^1$, by performing this operation we get a manifold that is $\mathbf{S}^1$ times the two discs glued together along their boundaries. Therefore, with this surgery we obtain $\mathbf{S}^2 \times \mathbf{S}^1$, and (36) then gives

$$(43) \qquad Z(\mathbf{S}^2 \times \mathbf{S}^1) = \langle 0|0 \rangle = 1.$$

If we do the gluing, however, after performing an S-transformation on the $\mathbf{T}^2$ the resulting manifold is instead $\mathbf{S}^3$. To see this, notice that the complement to a solid torus inside $\mathbf{S}^3$ is indeed another solid torus whose non-contractible cycle is homologous to the contractible cycle in the first torus. We then find

$$(44) \qquad Z(\mathbf{S}^3) = \langle 0|S|0 \rangle = S_{00}.$$

By using Weyl's denominator formula,

$$(45) \qquad \sum_{w \in W} \epsilon(w) e^{w(\rho)} = \prod_{\alpha > 0} 2 \sinh \frac{\alpha}{2},$$

where $\alpha > 0$ are positive roots, one finds

$$(46) \qquad Z(\mathbf{S}^3) = \frac{1}{(k+y)^{r/2}} \left(\frac{\mathrm{Vol}\, \Lambda^w}{\mathrm{Vol}\, \Lambda^r} \right)^{\frac{1}{2}} \prod_{\alpha > 0} 2 \sin \left(\frac{\pi(\alpha \cdot \rho)}{k+y} \right).$$

The above result can be generalized in order to compute path integrals in $\mathbf{S}^3$ with some knots and links. Consider a solid torus where a Wilson line in representation R has been inserted. The corresponding state is $|R\rangle$, as we explained before. If we now glue this to an empty solid torus after an S-transformation, we obtain a trivial knot, or *unknot*, in $\mathbf{S}^3$. The path integral with the insertion is then,

$$(47) \qquad Z_R = \langle 0|S|R \rangle.$$

It follows that the normalized vacuum expectation value for the unknot in $\mathbf{S}^3$, in representation R, is given by

$$(48) \qquad W_R(\text{unknot}) = \frac{S_{0R}}{S_{00}} = \frac{\sum_{w \in W} \epsilon(w) \, e^{-\frac{2\pi i}{k+y} \rho \cdot w(\Lambda + \rho)}}{\sum_{w \in W} \epsilon(w) \, e^{-\frac{2\pi i}{k+y} \rho \cdot w(\rho)}}.$$

This expression can be written in terms of characters of the group G. Remember that the character of the representation R, evaluated on an element $a \in \Lambda^w \otimes \mathbf{R}$ is defined by

$$(49) \qquad \mathrm{ch}_R(a) = \sum_{\mu \in M_R} e^{a \cdot \mu},$$

where M_R is the set of weights associated to the irreducible representation R. By using Weyl's character formula we can write

(50)
$$W_R(\text{unknot}) = \text{ch}_R\left[-\frac{2\pi i}{k+y}\rho\right].$$

Moreover, using (45), we finally obtain

(51)
$$W_R(\text{unknot}) = \prod_{\alpha>0} \frac{\sin\left(\frac{\pi}{k+y}\alpha \cdot (\Lambda+\rho)\right)}{\sin\left(\frac{\pi}{k+y}\alpha \cdot \rho\right)}.$$

This quantity is often called the *quantum dimension* of R, and it is denoted by $\dim_q R$.

We can also consider a solid torus with a Wilson loop in representation R, glued to another solid torus with the representation R' through an S-transformation. What we obtain is clearly a link in $\mathbf{S}^3$ with two components, which is the Hopf link shown in Fig. 1. Carefully taking into account the orientation, we find that this is the Hopf link with linking number $+1$. The path integral with this insertion is:

(52)
$$Z_{RR'} = \langle R'|S|R\rangle,$$

so the normalized vacuum expectation value is

(53)
$$\mathcal{W}_{RR'} \equiv W_{RR'}(\text{Hopf}^{+1}) = \frac{S_{\overline{R}'R}}{S_{00}} = \frac{S_{R'R}^{-1}}{S_{00}},$$

where the superscript $+1$ refers to the linking number. Here, we have used that the bras $\langle R|$ are canonically associated to conjugate representations $\overline{R}$, and that $S_{\overline{R}'R} = S_{R'R}^{-1}$ (see for example Di Francesco et al. 1997). Therefore, the Chern–Simons invariant of the Hopf link is essentially an S-matrix element. In order to obtain the invariant of the Hopf link with linking number -1, we notice that the two Hopf links can be related by changing the orientation of one of the components. Since this is equivalent to conjugating the representation, we find

(54)
$$W_{RR'}(\text{Hopf}^{-1}) = \frac{S_{R'R}}{S_{00}}.$$

When we take $G = U(N)$, the above vacuum expectation values for unknots and Hopf links can be evaluated very explicitly in terms of Schur polynomials. It is well known that the character of the unitary group in the representation R is given by the Schur polynomial s_R (see for example Fulton and Harris 1991). There is a precise relation between the element a on which one evaluates the character in (49) and the variables entering the Schur polynomial. Let μ_i, $i = 1, \cdots, N$, be the weights associated to the fundamental representation of $U(N)$. Notice that, if R is given by a Young tableau whose rows have lengths $l_1 \geq \cdots \geq l_N$, then $\Lambda_R = \sum_i l_i \mu_i$. We also have

(55)
$$\rho = \sum_{i=1}^{N} \frac{1}{2}(N - 2i + 1)\mu_i.$$

Let $a \in \Lambda^w \otimes \mathbf{R}$ be given by

$$
(56) \qquad a = \sum_{i=1}^{N} a_i \mu_i.
$$

Then,

$$
(57) \qquad \mathrm{ch}_R[a] = s_R(x_i = e^{a_i}).
$$

For example, in the case of the quantum dimension, one has $\dim_q R = \dim_q \overline{R}$, and we find

$$
(58) \qquad \dim_q R = s_R(x_i = q^{\frac{1}{2}(N-2i+1)}),
$$

where q is given in (12). By using that s_R is homogeneous of degree $\ell(R)$ in the variables x_i we finally obtain

$$
\dim_q R = \lambda^{\ell(R)/2} s_R(x_i = q^{-i+\frac{1}{2}})
$$

where $\lambda = q^N$ as in (12), and there are N variables x_i. The quantum dimension can be written very explicitly in terms of the q-numbers:

$$
(59) \qquad [a] = q^{\frac{a}{2}} - q^{-\frac{a}{2}}, \quad [a]_\lambda = \lambda^{\frac{1}{2}} q^{\frac{a}{2}} - \lambda^{-\frac{1}{2}} q^{-\frac{a}{2}}.
$$

If R corresponds to a Young tableau with c_R rows of lengths l_i, $i = 1, \cdots, c_R$, the quantum dimension is given by:

$$
(60) \qquad \dim_q R = \prod_{1 \le i < j \le c_R} \frac{[l_i - l_j + j - i]}{[j - i]} \prod_{i=1}^{c_R} \frac{\prod_{v=-i+1}^{l_i-i} [v]_\lambda}{\prod_{v=1}^{l_i} [v - i + c_R]}.
$$

It is easy to check that in the limit $k + N \to \infty$ (i.e. in the semi-classical limit) the quantum dimension becomes the dimension of the representation R. Notice that the quantum dimension is a rational function of $q^{\pm\frac{1}{2}}$, $\lambda^{\pm\frac{1}{2}}$. This is a general property of all normalized vacuum expectation values of knots and links in $\mathbf{S}^3$.

The S-matrix elements that appear in (53) and (54) can be evaluated through the explicit expression (40), by using the relation between $U(N)$ characters and Schur functions that we explained above. Notice first that

$$
(61) \qquad \frac{S_{R_1 R_2}^{-1}}{S_{00}} = \mathrm{ch}_{R_1}\left[\frac{2\pi i}{k+y}(\Lambda_{R_2} + \rho)\right]\mathrm{ch}_{R_2}\left[\frac{2\pi i}{k+y}\rho\right].
$$

If we denote by $l_i^{R_2}$, $i = 1, \cdots, c_{R_2}$ the lengths of rows for the Young tableau corresponding to R_2, it is easy to see that

$$
(62) \qquad \mathcal{W}_{R_1 R_2}(q, \lambda) = (\lambda q)^{\frac{\ell(R_1)}{2}} s_{R_1}(x_i = q^{l_i^{R_2}-i}) \dim_q R_2,
$$

where we set $l_i^{R_2} = 0$ for $i > c_{R_2}$. A convenient way to evaluate $s_{R_1}(x_i = q^{l_i^R - i})$ for a partition $\{l_i^R\}_{\{i=1,\cdots,c_R\}}$ associated to R is to use the Jacobi–Trudi formula (188). It is easy to show that the generating functional of elementary symmetric functions (184) for this specialization is given by

$$(63) \qquad E_R(t) = E_\emptyset(t) \prod_{j=1}^{c_R} \frac{1 + q^{l_j^R - j} t}{1 + q^{-j} t},$$

where

$$(64) \qquad E_\emptyset(t) = 1 + \sum_{n=1}^\infty a_n t^n,$$

and the coefficients a_n are defined by

$$(65) \qquad a_n = \prod_{r=1}^n \frac{1 - \lambda^{-1} q^{r-1}}{q^r - 1}.$$

The formula (62), together with the expressions above for $E_R(t)$, provides an explicit expression for (53) as a rational function of $q^{\pm\frac{1}{2}}$, $\lambda^{\pm\frac{1}{2}}$, and it was first written down by Morton and Lukac (2003).

2.4 Framing Dependence

In the above discussion on the correlation functions of Wilson loops we have glossed over an important ingredient. We already mentioned that, in order to define the partition function of Chern–Simons theory at the quantum level, one has to specify a framing of the three-manifold. It turns out that the evaluation of correlation functions like (8) also involves a choice of framing of the knots, as discovered by Witten (1989).

A good starting point to understand the framing is to take Chern–Simons theory with gauge group $U(1)$. The Abelian Chern–Simons theory turns out to be extremely simple, since the cubic term in (2) drops out, and we are left with a Gaussian theory (Polyakov 1988). $U(1)$ representations are labelled by integers, and the correlation function (8) can be computed exactly. In order to do that, however, one has to choose a framing for each of the knots $\mathcal{K}_\alpha$. This arises as follows: in evaluating the correlation function, contractions of the holonomies corresponding to different $\mathcal{K}_\alpha$ produce the following integral:

$$(66) \qquad \mathrm{lk}(\mathcal{K}_\alpha, \mathcal{K}_\beta) = \frac{1}{4\pi} \oint_{\mathcal{K}_\alpha} dx^\mu \oint_{\mathcal{K}_\beta} dy^\nu \epsilon_{\mu\nu\rho} \frac{(x-y)^\rho}{|x-y|^3}.$$

This is a topological invariant, i.e. it is invariant under deformations of the knots $\mathcal{K}_\alpha$, $\mathcal{K}_\beta$, and it is, in fact, the Gauss integral representation of their linking number $\mathrm{lk}(\mathcal{K}_\alpha, \mathcal{K}_\beta)$ defined in (9). On the other hand, contractions of the holonomies corresponding to the same knot $\mathcal{K}$ involve the integral

$$(67) \qquad \phi(\mathcal{K}) = \frac{1}{4\pi} \oint_{\mathcal{K}} dx^\mu \oint_{\mathcal{K}} dy^\nu \epsilon_{\mu\nu\rho} \frac{(x-y)^\rho}{|x-y|^3}.$$

This integral is well defined and finite (see, for example, Guadagnini et al. 1990), and it is called the *cotorsion* or *writhe* of $\mathcal{K}$. It gives the self-linking number of $\mathcal{K}$: if we project $\mathcal{K}$ on a plane, and we denote by $n_\pm(\mathcal{K})$ the number of positive (negative) crossings as indicated in Fig. 2, then we have that

$$(68) \qquad \phi(\mathcal{K}) = n_+(\mathcal{K}) - n_-(\mathcal{K}).$$

The problem is that the cotorsion is not invariant under deformations of the knot. In order to preserve topological invariance of the correlation function, one has to choose another definition of the composite operator $(\oint_{\mathcal{K}} A)^2$ by means of a framing. A framing of the knot consists of choosing another knot $\mathcal{K}^f$ around $\mathcal{K}$, specified by a normal vector field n. The cotorsion $\phi(\mathcal{K})$ then becomes

$$(69) \qquad \phi_f(\mathcal{K}) = \frac{1}{4\pi} \oint_{\mathcal{K}} dx^\mu \oint_{\mathcal{K}^f} dy^\nu \epsilon_{\mu\nu\rho} \frac{(x-y)^\rho}{|x-y|^3} = \mathrm{lk}(\mathcal{K}, \mathcal{K}^f).$$

The correlation function that we obtain in this way is a topological invariant (since it only involves linking numbers) but the price that we have to pay is that our regularization depends on a set of integers $p_\alpha = \mathrm{lk}(\mathcal{K}_\alpha, \mathcal{K}_\alpha^f)$ (one for each knot). The correlation function (8) can now be computed, after choosing the framings, as follows:

$$(70) \qquad \left\langle \prod_\alpha \exp\left(n_\alpha \oint_{\mathcal{K}_\alpha} A\right) \right\rangle = \exp\left\{ \frac{\pi i}{k}\left(\sum_\alpha n_\alpha^2 p_\alpha + \sum_{\alpha \neq \beta} n_\alpha n_\beta \, \mathrm{lk}(\mathcal{K}_\alpha, \mathcal{K}_\beta) \right) \right\}.$$

This regularization is simply the 'point-splitting' method familiar in the context of quantum field theory.

Let us now consider Chern–Simons theory with gauge group $U(N)$, and suppose that we are interested in the computation of (8), in the context of perturbation theory. It is easy to see that self-contractions of the holonomies lead to the same kind of ambiguities that we found in the Abelian case, i.e. a choice of framing has to be made for each knot $\mathcal{K}_\alpha$. The only difference from the Abelian case is that the self-contraction of $\mathcal{K}_\alpha$ gives a group factor $\mathrm{Tr}_{R_\alpha}(T_a T_a)$, where T_a is a basis of the Lie algebra (see, for example, Guadagnini et al. 1990). The precise result can be better stated as the effect on the correlation function (8) under a change of framing, and it says that, under a change of framing of $\mathcal{K}_\alpha$ by p_α units, the vacuum expectation value of the product of Wilson loops changes as follows (Witten 1989):

$$(71) \qquad W_{R_1 \cdots R_L} \to \exp\left[2\pi i \sum_{\alpha=1}^L p_\alpha h_{R_\alpha} \right] W_{R_1 \cdots R_L}.$$

In this equation, h_R is the conformal weight of the WZW primary field corresponding to the representation R. One can write (41) as

(72)
$$h_R = \frac{C_R}{2(k+N)},$$

where $C_R = \mathrm{Tr}_R(T_a T_a)$ is the quadratic Casimir in the representation R. For $U(N)$ one has

(73)
$$C_R = N\ell(R) + \kappa_R,$$

where $\ell(R)$ is the total number of boxes in the tableau, and

(74)
$$\kappa_R = \ell(R) + \sum_i \left(l_i^2 - 2il_i\right).$$

In terms of the variables (12) the change under framing (71) can be written as

(75)
$$W_{R_1\cdots R_L} \rightarrow q^{\frac{1}{2}\sum_{\alpha=1}^{L}\kappa_{R_\alpha}p_\alpha}\lambda^{\frac{1}{2}\sum_{\alpha=1}^{L}\ell(R_\alpha)p_\alpha}W_{R_1\cdots R_L}.$$

Therefore, the evaluation of vacuum expectation values of Wilson loop operators in Chern–Simons theory depends on a choice of framing for knots. It turns out that for knots and links in $\mathbf{S}^3$, there is a *standard* or canonical framing, defined by requiring that the self-linking number is zero. The expressions we have given before for the Chern–Simons invariant of the unknot and the Hopf link are all in the standard framing. Once the value of the invariant is known in the standard framing, the value in any other framing specified by non-zero integers p_α can be easily obtained from (71).

2.5 The $1/N$ Expansion in Chern–Simons Theory

As we explained above, the perturbative series of Chern–Simons theory around the trivial connection can be re-expressed in terms of fatgraphs. In particular, one should be able to study the free energy of Chern–Simons theory on the three-sphere in the $1/N$ expansion, i.e. to expand it as in (30) and to resum all fatgraphs of fixed genus in this expansion to obtain the quantities $F_g(t)$. In this section we will obtain closed expressions for $F_{g,h}$ and $F_g(t)$ in the case of Chern–Simons theory defined on $\mathbf{S}^3$, following Gopakumar and Vafa (1998a, 1999). For earlier work on the $1/N$ expansion of Chern–Simons theory, see Camperi et al. (1990), Periwal (1993) and Correale and Guadagnini (1994).

A direct computation of $F_{g,h}$ from perturbation theory is difficult, since it involves the evaluation of integrals of products of propagators over the three-sphere. However, in the case of $\mathbf{S}^3$ we have an exact expression for the partition function and we can expand it in both x and N to obtain the coefficients of (30). The partition function of CS with gauge group $U(N)$ on the three-sphere can be obtained from the formula (46) for $SU(N)$ after multiplying it by an overall $N^{1/2}/(k+N)^{1/2}$, which is the partition function of the $U(1)$ factor. The final result is

(76)
$$Z = \frac{1}{(k+N)^{N/2}} \prod_{\alpha>0} 2\sin\left(\frac{\pi(\alpha \cdot \rho)}{k+N}\right).$$

Using the explicit description of the positive roots of $SU(N)$, one gets

(77)
$$F = \log Z = -\frac{N}{2}\log(k+N) + \sum_{j=1}^{N-1}(N-j)\log\left[2\sin\frac{\pi j}{k+N}\right].$$

We can now write the sin as

(78)
$$\sin \pi z = \pi z \prod_{n=1}^{\infty}\left(1 - \frac{z^2}{n^2}\right),$$

and we find that the free energy is the sum of two parts. We will call the first one the *non-perturbative* part:

(79)
$$F^{\mathrm{np}} = -\frac{N^2}{2}\log(k+N) + \frac{1}{2}N(N-1)\log 2\pi + \sum_{j=1}^{N-1}(N-j)\log j,$$

and the other part will be called the *perturbative* part:

(80)
$$F^{\mathrm{p}} = \sum_{j=1}^{N}(N-j)\sum_{n=1}^{\infty}\log\left[1 - \frac{j^2 g_s^2}{4\pi^2 n^2}\right],$$

where we have denoted

(81)
$$g_s = \frac{2\pi}{k+N},$$

which, as we will see later, coincides with the open string coupling constant under the gauge/string theory duality.

To see that (79) has a non-perturbative origin, we rewrite it as

(82)
$$F^{\mathrm{np}} = \log \frac{(2\pi g_s)^{\frac{1}{2}N^2}}{\mathrm{vol}(U(N))},$$

where we used the explicit formula

(83)
$$\mathrm{vol}(U(N)) = \frac{(2\pi)^{\frac{1}{2}N(N+1)}}{G_2(N+1)},$$

and $G_2(N)$ is Barnes function. This indeed corresponds to the volume of the gauge group in the one-loop contribution (17), where $A^{(c)}$ is in this case the trivial flat connection. Therefore, F^{np} is the log of the prefactor of the path integral, which is not captured by Feynman diagrams.

Let us now work out the perturbative part (80), following Gopakumar and Vafa (1998a, 1999). By expanding the log, using that $\sum_{n=1}^{\infty} n^{-2k} = \zeta(2k)$, and the formula

$$(84) \qquad \sum_{j=1}^{N} j^k = \frac{1}{k+1} \sum_{l=1}^{k+1} (-1)^{k-l+1} \binom{k+1}{l} B_{k+1-l} N^l,$$

where B_n are Bernoulli numbers, we find that (80) can be written as

$$(85) \qquad F^{\mathrm{P}} = \sum_{g=0}^{\infty} \sum_{h=2}^{\infty} F_{g,h}^{\mathrm{P}} g_s^{2g-2+h} N^h,$$

where $F_{g,h}^{\mathrm{P}}$ is given by:

$$F_{0,h}^{\mathrm{P}} = -\frac{|B_{h-2}|}{(h-2)h!}, \quad h \geq 4,$$

$$(86) \qquad F_{1,h}^{\mathrm{P}} = \frac{1}{12} \frac{|B_h|}{h\, h!}.$$

Notice that $F_{0,h}^{\mathrm{P}}$ vanishes for $h \leq 3$. For $g \geq 2$ one obtains

$$(87) \qquad F_{g,h}^{\mathrm{P}} = \frac{\zeta(2g-2+h)}{(2\pi)^{2g-3+h}} \binom{2g-3+h}{h} \frac{B_{2g}}{2g(2g-2)}.$$

This gives the contribution of connected diagrams with two loops and beyond to the free energy of Chern–Simons theory on the sphere. The nonperturbative part also admits an asymptotic expansion that can be easily worked out by expanding the Barnes function that appears in the volume factor (Periwal 1993; Ooguri and Vafa 2002). One gets:

$$(88) \quad F^{\mathrm{np}} = \frac{N^2}{2} \left(\log(N g_s) - \frac{3}{2} \right) - \frac{1}{12} \log N + \zeta'(-1) + \sum_{g=2}^{\infty} \frac{B_{2g}}{2g(2g-2)} N^{2-2g}.$$

In order to find $F_g(t)$ we have to sum over the holes, as in (33). The 't Hooft parameter is given by $t = xN = i g_s N$, and

$$(89) \qquad F_g^{\mathrm{P}}(t) = \sum_{h=1}^{\infty} F_{g,h}^{\mathrm{P}}(-it)^h.$$

Let us first focus on $g \geq 2$. To perform the sum explicitly, we again write the ζ function as $\zeta(2g-2+2p) = \sum_{n=1}^{\infty} n^{2-2g-2p}$, and use the binomial series,

$$(90) \qquad \frac{1}{(1-z)^q} = \sum_{n=0}^{\infty} \binom{q+n-1}{n} z^n.$$

to obtain:

$$(91) \qquad F_g^{\mathrm{p}}(t) = \frac{(-1)^g |B_{2g} B_{2g-2}|}{2g(2g-2)(2g-2)!} + \frac{B_{2g}}{2g(2g-2)} \sum_{n\in\mathbf{Z}}{}' \frac{1}{(-it+2\pi n)^{2g-2}},$$

where $'$ means that we omit $n = 0$. Now we notice that, if we write

$$(92) \qquad F^{\mathrm{np}} = \sum_{g=0}^{\infty} F_g^{\mathrm{np}}(t) g_s^{2g-2},$$

then for, $g \geq 2$, $F_g^{\mathrm{np}}(t) = B_{2g}/(2g(2g-2)(-it)^{2g-2}$, which is precisely the $n = 0$ term missing in (91). We then define:

$$(93) \qquad F_g(t) = F_g^{\mathrm{p}}(t) + F_g^{\mathrm{np}}(t).$$

Finally, since

$$(94) \qquad \sum_{n\in\mathbf{Z}} \frac{1}{n+z} = \frac{2\pi i}{1 - e^{-2\pi i z}},$$

by taking derivatives w.r.t. z we can write

$$(95) \qquad F_g(t) = \frac{(-1)^g |B_{2g} B_{2g-2}|}{2g(2g-2)(2g-2)!} + \frac{|B_{2g}|}{2g(2g-2)!} \mathrm{Li}_{3-2g}(e^{-t}),$$

again for $g \geq 2$. The function Li_j appearing in this equation is the polylogarithm of index j, defined by

$$(96) \qquad \mathrm{Li}_j(x) = \sum_{n=1}^{\infty} \frac{x^n}{n^j}.$$

The computation for $g = 0, 1$ is very similar, and one obtains:

$$(97) \qquad \begin{aligned} F_0(t) &= -\frac{t^3}{12} + \frac{\pi^2 t}{6} + \zeta(3) + \mathrm{Li}_3(e^{-t}), \\ F_1(t) &= \frac{t}{24} + \frac{1}{12}\log(1 - e^{-t}). \end{aligned}$$

This gives the resummed functions $F_g(t)$ introduced in (33) for all $g \geq 0$.

3 Topological Strings

In this section we give a rough presentation of Gromov–Witten invariants. Detailed definitions and constructions can be found for example in Cox and Katz (1999).

3.1 Topological Strings and Gromov–Witten Invariants

In order to define Gromov–Witten invariants, the starting point is the moduli space of possible metrics (or equivalently, complex structures) on a Riemann surface with punctures, which is the famous Deligne-Mumford space $\overline{M}_{g,n}$ of n-pointed stable curves (the definition of what stable means can be found for example in Harris and Morrison 1998). Let X be a Kähler manifold. The relevant moduli space in Gromov–Witten theory is denoted by

$$(98) \qquad \overline{M}_{g,n}(X,\beta)$$

where $\beta \in H_2(X)$. This is a generalization of $\overline{M}_{g,n}$, and depends on a choice of a two-homology class β in X. Very roughly, a point in $\overline{M}_{g,n}(X,\beta)$ can be written as $(f, \Sigma_g, p_1, \cdots, p_n)$ and is given by (a) a point in $\overline{M}_{g,n}$, i.e. a Riemann surface with n punctures, $(\Sigma_g, p_1, \cdots, p_n)$, together with a choice of complex structure on Σ_g, and (b) a map $f : \Sigma_g \to X$ that is holomorphic with respect to this choice of complex structure and such that $f_*[\Sigma_g] = \beta$. The set of all such points forms a good moduli space provided a certain number of conditions are satisfied (see for example Cox and Katz (1999) and Hori et al. (2003) for a detailed discussion of these issues). $\overline{M}_{g,n}(X,\beta)$ is the basic moduli space we will need in the theory of topological strings. Its complex virtual dimension is given by

$$(99) \qquad (1-g)(d-3) + n + \int_{\Sigma_g} f^*(c_1(X)).$$

We also have two natural maps

$$(100) \qquad \begin{aligned} \pi_1 &: \overline{M}_{g,n}(X,\beta) \longrightarrow X^n, \\ \pi_2 &: \overline{M}_{g,n}(X,\beta) \longrightarrow \overline{M}_{g,n}. \end{aligned}$$

The first map is easy to define: given a point $(f, \Sigma_g, p_1, \cdots, p_n)$ in $\overline{M}_{g,n}(X,\beta)$, we just compute $(f(p_1), \cdots, f(p_n))$. The second map essentially sends $(f, \Sigma_g, p_1, \cdots, p_n)$ to $(\Sigma_g, p_1, \cdots, p_n)$, i.e. forgets the information about the map and keeps the information about the punctured curve.

We can now formally define the Gromov–Witten invariant $I_{g,n,\beta}$ as follows. Let us consider cohomology classes $\phi_1, \cdots, \phi_n$ in $H^*(X)$. If we pull back their tensor product to $H^*(\overline{M}_{g,n}(X,\beta))$ via π_1, we get a differential form on the moduli space of maps that we can integrate (as long as there is a well-defined fundamental class for this space):

$$(101) \qquad I_{g,n,\beta}(\phi_1, \cdots, \phi_n) = \int_{\overline{M}_{g,n}(X,\beta)} \pi_1^*(\phi_1 \otimes \cdots \otimes \phi_n).$$

The Gromov–Witten invariant $I_{g,n,\beta}(\phi_1, \cdots, \phi_n)$ vanishes unless the degree of the form equals the dimension of the moduli space. Therefore, we have the following constraint:

$$(102) \qquad \frac{1}{2} \sum_{i=1}^{n} \deg(\phi_i) = (1-g)(d-3) + n + \int_{\Sigma_g} f^*(c_1(X)).$$

Notice that Calabi–Yau threefolds play a special role in the theory, since for those targets the virtual dimension only depends on the number of punctures, and therefore the above condition is always satisfied if the forms ϕ_i have degree 2.

When $n = 0$, one gets an invariant $I_{g,0,\beta}$ that does not require any insertions. This is the Gromov–Witten invariant on which we will focus, and we will denote it by $N_{g,\beta}$. Notice that these invariants are in general *rational*, due to the orbifold character of the moduli spaces involved. It is very convenient to introduce the generating functional of these invariants at fixed genus. This is defined as follows. First, choose a basis $[\Sigma_i] \in H_2(X)$ in such a way that

$$(103) \qquad \beta = \sum_{i=1}^{h^{1,1}(X)} \beta_i [\Sigma_i].$$

We also introduce $h^{1,1}(X)$ *complexified Kähler parameters* t_i. They are defined as

$$(104) \qquad t_i = \int_{\Sigma_i} \omega.$$

In this equation, ω is the complexified Kähler class,

$$(105) \qquad \omega = J + iB,$$

where J is the Kähler class and B is the B-field. Finally, we introduce

$$(106) \qquad \beta \cdot t = \sum_{i=1}^{h^{1,1}(X)} \beta_i t_i = \int_{\beta} \omega.$$

With these ingredients, we define the *topological string amplitude at genus g* as the generating functional

$$(107) \qquad F_g(t) = \sum_{\beta \in H_2(X)} N_{g,\beta} e^{-\beta \cdot t}.$$

The *total topological string amplitude* sums this to all genera,

$$(108) \qquad F(g_s, t) = \sum_{g=0}^{\infty} F_g(t) g_s^{2g-2}.$$

It is also convenient to consider the exponentiated functional, which is called the *topological string partition function*,

$$(109) \qquad Z(g_s, t) = \exp F(g_s, t).$$

An important goal in Gromov–Witten theory is to provide effective tools for the computation of these quantities. The main reason why physics is useful in doing this is because the $F_g(t)$ are couplings in type II string theory, and can be also obtained as free energies of topological string theory, a topological version of string theory which is obtained by coupling topological sigma models to topological gravity (hence the name of topological string quantities for these quantities). For an exposition of some of the relevant physics background, see Mariño (2005).

3.2 Integrality Properties and Gopakumar–Vafa Invariants

It was shown by Gopakumar and Vafa (1998b) that the total free energy $F(g_s, t)$ can be expressed in terms of *integer numbers* n_β^g as follows

$$(110) \qquad F(g_s, t) = \sum_{g=0}^{\infty} \sum_{\beta} \sum_{d=1}^{\infty} n_\beta^g \frac{1}{d} \left(2 \sin \frac{d g_s}{2} \right)^{2g-2} Q^{d\beta}.$$

The integers n_β^g are known as *Gopakumar–Vafa invariants*. They are true invariants of the Calabi–Yau manifold X, in the sense that they do not depend on smooth deformations of the target geometry, This is in contrast to the quantities $n_\beta^{(j_L, j_R)}$, which do depend on deformations. As usual, by tracing over a non-invariant quantity with signs we obtain an invariant quantity.

The structure result (110) implies that Gromov–Witten invariants of closed strings, which are in general rational, can be written in terms of these integer invariants. In fact, by knowing the Gromov–Witten invariants $N_{g,\beta}$ we can explicitly compute the Gopakumar–Vafa invariants from (110) (an explicit inversion formula can be found in Bryan and Pandharipande 2001). By expanding in g_s, it is easy to show that the Gopakumar–Vafa formula (110) predicts the following expression for $F_g(t)$:

(111)
$$F_g(t) = \sum_{\beta} \left(\frac{|B_{2g}| n_\beta^0}{2g(2g-2)!} + \frac{2(-1)^g n_\beta^2}{(2g-2)!} \pm \cdots - \frac{g-2}{12} n_\beta^{g-1} + n_\beta^g \right) \mathrm{Li}_{3-2g}(Q^\beta),$$

where Li_j is the polylogarithm defined in (96). The appearance of the polylogarithm of order $3 - 2g$ in F_g was first predicted from type IIA/heterotic string duality by Mariño and Moore (1999).

The structure found by Gopakumar and Vafa solves some longstanding issues in the theory of Gromov–Witten invariants, in particular the enumerative meaning of the invariants. Two obstructions to finding obvious enumerative meaning to Gromov–Witten invariants are *multicovering* and *bubbling*. Multicovering arises as follows. Suppose one finds a holomorphic map $x : \mathbb{P}^1 \to X$ in genus zero and in the class β. Then, simply by composing this with a degree d cover $\mathbb{P}^1 \to \mathbb{P}^1$, one can find another holomorphic map in the class $d\beta$. Therefore, at every degree, in order to count the actual number of 'primitive' holomorphic curves, one should subtract from the corresponding Gromov–Witten invariant the contributions coming from multicovering of curves with lower degree. Another geometric effect that has to be taken into account is *bubbling* (see, for example, Bershadsky et al. 1993, 1994). Imagine that one finds a map $x : \Sigma_g \to X$ from a genus g Riemann surface to a Calabi–Yau threefold. By gluing to Σ_g a small Riemann surface of genus h, and making it very small, one can find an approximate holomorphic map from a Riemann surface whose genus is topologically $g+h$. This means that 'primitive' maps at genus g contribute to all genera $g' > g$, and in order to count curves properly one should take this effect into account.

The formula (111) gives a precise answer to these questions. Consider, for example, the structure of F_0. According to the above formula, the contribution of a Gopakumar–Vafa invariant is given by the function Li_3:

$$(112) \qquad \sum_{d=1}^{\infty} \frac{Q^{d\beta}}{d^3}.$$

This gives the contribution of all the multicoverings of a given 'primitive' curve, where d is the degree of the multicovering. In addition, it says that each cover has a weight $1/d^3$. Therefore, the invariant n_β^0 corresponds to primitive holomorphic maps, and the non-integrality of genus-zero Gromov–Witten invariants is due to the effects of multicovering. The multicovering phenomenon in genus 0 was found experimentally in Candelas et al. (1991) and later derived in the context of Gromov–Witten theory by Aspinwall and Morrison (1993). The structure result of Gopakumar and Vafa also predicts that the multicovering of degree d of a genus g curve contributes with a weight d^{3-2g} (coming from Li_{3-2g}). Moreover, the formula (111) implies that a genus $h < g$ Gopakumar–Vafa invariant contributes to $F_g(t)$ with a precise weight, and this corresponds to the bubbling effects we mentioned before. For example, a genus 0 Gopakumar–Vafa invariant contributes to F_g with a weight $|B_{2g}|/(2g(2g-2)!)$.

3.3 Open Topological Strings

So far we have discussed the Gromov–Witten theory for the case of closed Riemann surfaces, but the theory can be (at least formally) extended to the open case. The natural starting point is to consider maps from a Riemann surface $\Sigma_{g,h}$ of genus g with h holes. Such models were analysed in detail by Witten (1995). The main issue is, of course, to specify boundary conditions for the maps $f : \Sigma_{g,h} \to X$. It turns out that the relevant boundary conditions are Dirichlet and given by Lagrangian submanifolds of the Calabi–Yau X. A Lagrangian submanifold $\mathcal{L}$ is a cycle on which the Kähler form vanishes:

$$(113) \qquad\qquad J|_{\mathcal{L}} = 0.$$

If we denote by C_i, $i = 1, \cdots, h$, the boundaries of $\Sigma_{g,h}$ we have to pick a Lagrangian submanifold $\mathcal{L}$, and consider holomorphic maps such that

$$(114) \qquad\qquad f(C_i) \subset \mathcal{L}.$$

Once boundary conditions have been specified, we look at holomorphic maps from open Riemann surfaces of genus g and with h holes to the Calabi–Yau X, with Dirichlet boundary conditions specified by $\mathcal{L}$. These holomorphic maps are called *open string instantons*, and can also be classified topologically. The topological sector of an open string instanton is given by two different kinds of data: the boundary part and the bulk part. For the bulk part, the topological sector is labelled by relative homology classes, since we are requiring the boundaries of $f_*[\Sigma_{g,h}]$ to end on $\mathcal{L}$. Therefore, we will set

$$(115) \qquad\qquad f_*[\Sigma_{g,h}] = \beta \in H_2(X, \mathcal{L}).$$

To specify the topological sector of the boundary, we will assume that $b_1(\mathcal{L}) = 1$, so that $H_1(\mathcal{L})$ is generated by a non-trivial one-cycle γ. We then have

$$(116) \qquad\qquad f_*[C_i] = w_i \gamma, \quad w_i \in \mathbf{Z}, \quad i = 1, \cdots, h,$$

in other words, w_i is the winding number associated to the map f restricted to C_i. We will collect these integers into a single h-uple denoted by $w = (w_1, \cdots, w_h)$.

The free energy of open topological string theory at fixed genus and boundary data w, which we denote by $F_{w,g}(t)$, can be computed as a sum over open string instantons labelled by the bulk classes:

$$(117) \qquad\qquad F_{w,g}(t) = \sum_{\beta} F_{w,g,\beta}\, e^{-\beta \cdot t}.$$

In this equation, the sum is over relative homology classes $\beta \in H_2(X, \mathcal{L})$. The quantities $F_{w,g,\beta}$ are *open Gromov–Witten invariants*. They 'count' in an appropriate sense the number of holomorphically embedded Riemann surfaces of genus g in X with Lagrangian boundary conditions specified by $\mathcal{L}$, and in the class represented by β, w. They are in general rational numbers. In contrast to conventional Gromov–Witten invariants, a rigorous theory of open Gromov–Witten invariants is not yet available. However, localization techniques make it possible to compute them in some situations (Katz and Liu 2002; Li and Song 2002; Graber and Zaslow 2002; Mayr 2002).

In order to consider all topological sectors, we have to introduce the string coupling constant g_s, which takes care of the genus, as well as a Hermitian $M \times M$ matrix V, which takes care of the different winding numbers w. The total free energy is defined by

$$(118) \qquad F(V) = \sum_{g=0}^{\infty} \sum_{h=1}^{\infty} \sum_{w_1, \cdots, w_h} \frac{i^h}{h!} g_s^{2g-2+h} F_{w,g}(t) \operatorname{Tr} V^{w_1} \cdots \operatorname{Tr} V^{w_h}.$$

The factor i^h is introduced for convenience, while $h!$ is a symmetry factor which takes into account that the holes are indistinguishable. Notice that, in order to distinguish all possible topological sectors, one has to take V to have infinite rank, and formally we can think about the different traces in (118) as symmetric functions in an infinite number of variables.

If the winding numbers w_i in (118) are all positive, the product of traces of V in (118) can be written in terms of $\mathrm{Tr}_R V$ for representations R with a small number of boxes:

$$(119) \qquad F(V) = \sum_R F_R(g_s, t) \mathrm{Tr}_R V,$$

Negative winding numbers can be introduced through another set of representations. We have also assumed that the boundary conditions are specified by a single Lagrangian submanifold with a single non-trivial one-cycle. When there are more one-cycles in the geometry, say L, providing possible boundary conditions for the open strings, the above formalism has to be generalized in an obvious way: one needs to specify L sets of winding numbers $w^{(\alpha)}$, and the generating functional (119) depends on L different matrices V_α, $\alpha = 1, \cdots, L$. The total partition function is the formal exponential of the total free energy and it has the structure

$$(120) \qquad Z(V_i) = \sum_{R_1, \cdots, R_{2L}} Z_{R_1 \cdots R_{2L}}(g_s, t) \prod_{\alpha=1}^{2L} \mathrm{Tr}_{R_\alpha} V_\alpha,$$

where the $R_{2\alpha-1}$, $R_{2\alpha}$ correspond to positive and negative winding numbers, respectively, for the α-th cycle.

4 Toric Geometry and Calabi–Yau Threefolds

4.1 Non-Compact Calabi–Yau Geometries: An Introduction

One of the main insights in the study of Gromov–Witten theory on Calabi–Yau threefolds is that the simplest models to study are associated to non-compact Calabi–Yau geometries based on manifolds of lower dimension. To construct these geometries, we start with complex manifolds in one or two complex dimensions, which in general will have a non-zero first Chern class. We then consider vector bundles over them (with the appropriate rank and curvature) that lead to a total three-dimensional space with zero first Chern class. In this way, we obtain Calabi–Yau threefolds whose non-trivial geometry is encoded in a lower-dimensional manifold, and therefore they are easier to study.

Let us first consider non-compact Calabi–Yau manifolds whose building block is a one-dimensional compact manifold. These manifolds will be given by a Riemann surface together with an appropriate bundle over it, and geometrically they can be regarded as the local geometry of an embedded Riemann surface in a general Calabi–Yau space. Indeed, consider a Riemann surface Σ_g holomorphically embedded inside a Calabi–Yau threefold X, and let us look at the holomorphic tangent bundle of X restricted to Σ_g. We have

$$\text{(121)} \qquad TX|_{\Sigma_g} = T\Sigma_g \oplus N_{\Sigma_g},$$

where N_{Σ_g} is a holomorphic rank-two complex vector bundle over Σ_g, called the normal bundle of Σ_g, and the Calabi–Yau condition $c_1(X) = 0$ gives

$$\text{(122)} \qquad c_1(N_{\Sigma_g}) = 2g - 2.$$

The Calabi–Yau X 'near Σ_g' then looks like the total space of the bundle

$$\text{(123)} \qquad N \to \Sigma_g,$$

where N is regarded here as a rank-two bundle over Σ_g satisfying (122). The non-compact space (123) is an example of a *local* Calabi–Yau threefold.

When $g = 0$ and $\Sigma_g = \mathbb{P}^1$ it is possible to be more precise about the bundle N. A theorem due to Grothendieck says that any holomorphic bundle over $\mathbb{P}^1$ splits into a direct sum of line bundles (for a proof, see for example Griffiths and Harris 1977, pp. 516–7). Line bundles over $\mathbb{P}^1$ are all of the form $\mathcal{O}(n)$, where $n \in \mathbf{Z}$. The bundle $\mathcal{O}(n)$ can be easily described in terms of two charts on $\mathbb{P}^1$: the north-pole chart, with co-ordinates z, Φ for the base and the fibre, respectively, and the south-pole chart, with co-ordinates z', Φ'. The change of co-ordinates is given by

$$\text{(124)} \qquad z' = 1/z, \quad \Phi' = z^{-n}\Phi.$$

We also have that $c_1(\mathcal{O}(n)) = n$. We then find that local Calabi–Yau manifolds that are made out of a two-sphere together with a bundle over it are all of the form

$$\text{(125)} \qquad \mathcal{O}(-a) \oplus \mathcal{O}(a - 2) \to \mathbb{P}^1,$$

since the degrees of the bundles have to sum up to -2 due to (122). An important case occurs when $a = 1$. The resulting non-compact manifold,

$$\text{(126)} \qquad \mathcal{O}(-1) \oplus \mathcal{O}(-1) \to \mathbb{P}^1,$$

is called the *resolved conifold* for reasons that will be explained later.

We can also consider non-compact Calabi–Yau threefolds based on compact complex surfaces. Consider a complex surface S embedded in a Calabi–Yau manifold X. As before, we can split the tangent bundle as

$$\text{(127)} \qquad TX|_S = TS \oplus N_S,$$

where the normal bundle N_S is now of rank one. The Calabi–Yau condition leads to

(128) $$c_1(\mathcal{N}_S) = c_1(K_S),$$

where K_S is the canonical line bundle over S, and we used that $c_1(TS) = -c_1(K_S)$. Therefore, we have $\mathcal{N}_S = K_S$. The Calabi–Yau X 'near S' looks like the total space of the bundle

(129) $$K_S \to S.$$

This construction gives a whole family of non-compact Calabi–Yau manifolds that are also referred to as local Calabi–Yau manifolds. A well-known example is $S = \mathbb{P}^2$, the two-dimensional projective space, which leads to the Calabi–Yau manifold

(130) $$\mathcal{O}(-3) \to \mathbb{P}^2,$$

also known as *local* $\mathbb{P}^2$. Another important example is $S = \mathbb{P}^1 \times \mathbb{P}^1$, which leads to *local* $\mathbb{P}^1 \times \mathbb{P}^1$.

4.2 Constructing Toric Calabi–Yau Manifolds

Many of the examples of non-compact Calabi–Yau threefolds considered above are *toric*, i.e. they have the structure of a torus fibration, and can be constructed in a systematic way by a 'cut and paste' procedure. In this section we will develop these techniques, following the approach of Aganagic et al. (2005).

$\mathbf{C}^3$

The elementary building block for the technique we want to develop is a very simple non-compact Calabi–Yau threefold, namely $\mathbf{C}^3$. We will now exhibit its structure as a $\mathbf{T}^2 \times \mathbb{R}$ fibration over $\mathbb{R}^3$, and we will encode this information in a simple trivalent, planar graph.

Let z_i be complex co-ordinates on $\mathbf{C}^3$, $i = 1, 2, 3$. We introduce three functions or Hamiltonians

$$r_\alpha(z) = |z_1|^2 - |z_3|^2,$$
$$r_\beta(z) = |z_2|^2 - |z_3|^2,$$
(131) $$r_\gamma(z) = \mathrm{Im}(z_1 z_2 z_3).$$

These Hamiltonians generate three flows on $\mathbf{C}^3$ via the standard symplectic form $\omega = i \sum_j dz_j \wedge d\bar{z}_j$ on $\mathbf{C}^3$ and the Poisson brackets

(132) $$\partial_v z_i = \{r_v, z_i\}_\omega, \quad v = \alpha, \beta, \gamma.$$

This gives the fibration structure that we were looking for: the base of the fibration, $\mathbb{R}^3$, is parameterized by the Hamiltonians (131), while the fibre $\mathbf{T}^2 \times$

$\mathbb{R}$ is parameterized by the flows associated to the Hamiltonians. In particular, the $\mathbf{T}^2$ fibre is generated by the circle actions

$$(133) \qquad e^{\alpha r_\alpha + \beta r_\beta} : \quad (z_1, z_2, z_3) \rightarrow (e^{i\alpha} z_1, e^{i\beta} z_2, e^{-i(\alpha+\beta)} z_3),$$

while r_γ generates the real line $\mathbb{R}$. We will call the cycle generated by r_α the $(0,1)$ cycle, and the cycle generated by r_β the $(1,0)$ cycle.

Notice that the $(0,1)$ cycle degenerates over the subspace of $\mathbf{C}^3$ described by $z_1 = 0 = z_3$, which is the subspace of the base $\mathbb{R}^3$ given by $r_\alpha = r_\gamma = 0$, $r_\beta \geq 0$. Similarly, over $z_2 = 0 = z_3$ the $(1,0)$-cycle degenerates over the subspace $r_\beta = r_\gamma = 0$ and $r_\alpha \geq 0$. Finally, the one-cycle parameterized by $\alpha + \beta$ degenerates over $z_1 = 0 = z_2$, where $r_\alpha - r_\beta = 0 = r_\gamma$ and $r_\alpha \leq 0$.

We will represent the $\mathbf{C}^3$ geometry by a graph that encodes the degeneration loci in the $\mathbb{R}^3$ base. In fact, it is useful to have a planar graph by taking $r_\gamma = 0$ and drawing the lines in the $r_\alpha - r_\beta$ plane. The degeneration locus will then be straight lines described by the equation $pr_\alpha + qr_\beta = \text{const.}$ Over this line the $(-q, p)$ cycle of the $\mathbf{T}^2$ degenerates. Therefore we correlate the degenerating cycles unambiguously with the lines in the graph (up to $(q, p) \rightarrow (-q, -p)$). This yields the graph in Fig. 12, drawn in the $r_\gamma = 0$ plane.

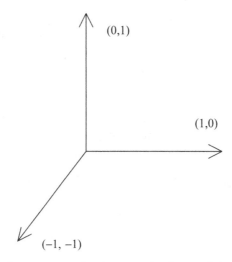

Fig. 12. This graph represents the degeneration locus of the $\mathbf{T}^2 \times \mathbb{R}$ fibration of $\mathbf{C}^3$ in the base $\mathbb{R}^3$ parameterized by $(r_\alpha, r_\beta, r_\gamma)$

There is a symmetry in the $\mathbf{C}^3$ geometry that makes it possible to find other representations by different toric graphs. These graphs are characterized by three vectors v_i that are obtained from those in Fig. 12 by an $SL(2, \mathbf{Z})$ transformation. The vectors have to satisfy

(134)
$$\sum_i v_i = 0.$$

The SL(2, **Z**) symmetry is inherited from the SL(2, **Z**) symmetry of **T**^{2} that appeared in Sect. 2 in the context of Chern–Simons theory. In the above discussion the generators $H_1(\mathbf{T}^2)$ have been chosen to be the one-cycles associated to r_α and r_β, but there are other choices that differ from this one by an SL(2, **Z**) transformation on the **T**^{2}. For example, we can choose r_α to generate a (p, q) one-cycle and r_β a (t, s) one-cycle, provided that $ps - qt = 1$. These different choices give different trivalent graphs. As we will see in the examples below, the construction of general toric geometries requires these more general graphs representing **C**^{3}.

The General Case

The non-compact, toric Calabi–Yau threefolds that we will study can be described as symplectic quotients. Let us consider the complex linear space **C**$^{N+3}$, described by $N + 3$ co-ordinates $z_1, \cdots, z_{N+3}$, and let us introduce N real equations of the form

(135)
$$\mu_A = \sum_{j=1}^{N+3} Q_A^j |z_j|^2 = t_A, \qquad A = 1, \cdots, N.$$

In this equation, Q_A^j are integers satisfying

(136)
$$\sum_{j=1}^{N+3} Q_A^j = 0.$$

This condition is equivalent to $c_1(X) = 0$, i.e. to the Calabi–Yau condition. We consider the action of the group $G_N = U(1)^N$ on the zs where the A-th $U(1)$ acts on z_j by
$$z_j \rightarrow \exp(iQ_A^j \alpha_A) z_j.$$
The space defined by (135), quotiented by the group action G_N,

(137)
$$X = \bigcap_{A=1}^{N} \mu_A^{-1}(t_A)/G_N$$

turns out to be a Calabi–Yau manifold (it can be seen that the condition (136) is equivalent to the Calabi–Yau condition). The N parameters t_A are Kähler moduli of the Calabi–Yau. This mathematical description of X appears in the study of the two-dimensional linear sigma model with $\mathcal{N} = (2, 2)$ supersymmetry (Witten 1993). The theory has $N + 3$ chiral fields, whose lowest components are the zs and are charged under N vector multiplets with charges Q_A^j. The equations (135) are the D-term equations, and after dividing by the $U(1)^N$ gauge group we obtain the Higgs branch of the theory.

The Calabi–Yau manifold X defined in (137) can be described by $\mathbf{C}^3$ geometries glued together in an appropriate way. Since each of these $\mathbf{C}^3$s is represented by the trivalent vertex depicted in Fig. 12, we will be able to encode the geometry of (137) into a trivalent graph. In order to provide this description, we must first find a decomposition of the set of all co-ordinates $\{z_j\}_{j=1}^{N+3}$ into triplets $U_a = (z_{i_a}, z_{j_a}, z_{k_a})$ that correspond to the decomposition of X into $\mathbf{C}^3$ patches. We pick one of the patches and we associate to it two Hamiltonians r_α, r_β as we did for $\mathbf{C}^3$ before. These two co-ordinates will be global co-ordinates in the base $\mathbb{R}^3$, therefore they will generate a globally defined $\mathbf{T}^2$ fibre. The third co-ordinate in the base is $r_\gamma = \text{Im}(\prod_{j=1}^{N+3} z_j)$, which is manifestly gauge invariant and moreover, patch by patch, can be identified with the co-ordinate used in the $\mathbf{C}^3$ example above. Equation (135) can then be used to find the action of $r_{\alpha,\beta}$ on the other patches.

We will now exemplify this procedure with two important examples: the resolved conifold and the local $\mathbb{P}^2$ geometry, which were introduced before as local Calabi–Yau geometries.

Example. *The resolved conifold.* The resolved conifold (126) has a description of the form (137), with $N = 1$. There is only one constraint given by

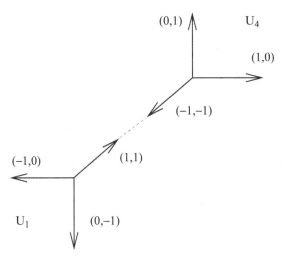

Fig. 13. The graph associated to the resolved conifold $\mathcal{O}(-1) \oplus \mathcal{O}(-1) \to \mathbb{P}^1$. This manifold is made out of two $\mathbf{C}^3$ patches glued through a common edge

$$(138) \qquad |z_1|^2 + |z_4|^2 - |z_2|^2 - |z_3|^2 = t,$$

and the $U(1)$ group acts as

$$(139) \qquad z_1, z_2, z_3, z_4 \to e^{i\alpha} z_1, e^{-i\alpha} z_2, e^{-i\alpha} z_3, e^{i\alpha} z_4.$$

Notice that, for $z_2 = z_3 = 0$, (138) describes a $\mathbb{P}^1$ whose area is proportional to t. Therefore, (z_1, z_4) can be taken as homogeneous co-ordinates of the $\mathbb{P}^1$ that is the basis of the fibration, while z_2, z_3 can be regarded as co-ordinates for the fibres.

Let us now give a description in terms of $\mathbf{C}^3$ patches glued together. The first patch will be defined by $z_4 \neq 0$. Using (138) we can solve for the modulus of z_4 in terms of the other co-ordinates, and using the $U(1)$ action we can gauge away its phase. Therefore, the patch will be parameterized by $U_4 = (z_1, z_2, z_3)$. The Hamiltonians will be, in this case,

$$
\begin{aligned}
r_\alpha(z) &= |z_2|^2 - |z_1|^2, \\
r_\beta(z) &= |z_3|^2 - |z_1|^2,
\end{aligned}
$$
(140)

which generate the actions

(141) $\qquad e^{\alpha r_\alpha + \beta r_\beta} : \quad (z_1, z_2, z_3) \; \rightarrow \; (e^{-i(\alpha+\beta)} z_1, e^{i\alpha} z_2, e^{i\beta} z_3).$

This patch will be represented by the same graph that we found for $\mathbf{C}^3$. The other patch will be defined by $z_1 \neq 0$, therefore we can write it as $U_1 = (z_4, z_2, z_3)$. However, in this patch z_1 is no longer a natural co-ordinate, but we can use (138) to rewrite the Hamiltonians as

$$
\begin{aligned}
r_\alpha(z) &= |z_4|^2 - |z_3|^2 - t, \\
r_\beta(z) &= |z_4|^2 - |z_2|^2 - t,
\end{aligned}
$$
(142)

generating the action

(143) $\qquad e^{\alpha r_\alpha + \beta r_\beta} : \quad (z_4, z_2, z_3) \; \rightarrow \; (e^{i(\alpha+\beta)} z_4, e^{-i\beta} z_2, e^{-i\alpha} z_3).$

The degeneration loci in this patch are the following: (1) $z_4 = 0 = z_2$, corresponding to the line $r_\beta = -t$ where a $(-1, 0)$ cycle degenerates; (2) $z_4 = 0 = z_3$, corresponding to the line $r_\alpha = -t$, and with a $(0, 1)$ cycle degenerating; (3) finally, $z_2 = 0 = z_3$, where $r_\alpha - r_\beta = 0$, and a $(1, 1)$ cycle degenerates. This patch is identical to the first one, and they are joined together through the common edge where $z_2 = 0 = z_3$. The full construction is represented in Fig. 13. Notice that the common edge of the graphs represents the $\mathbb{P}^1$ of the resolved conifold: along this edge, one of the $\mathbf{S}^1$s of $\mathbf{T}^2$ has degenerated, while the other only degenerates at the endpoints. An $\mathbf{S}^1$ fibration of an interval that degenerates at its endpoints is simply a two-sphere. The length of the edge is t, the Kähler parameter associated to the $\mathbb{P}^1$.

Example. *Local* $\mathbb{P}^2$. Let us now consider a more complicated example, namely local $\mathbb{P}^2$, which is the total space of the bundle (130). We can describe it again as in (137) with $N = 1$. There are four complex variables, $z_0, \cdots, z_3$, and the constraint (135) now reads

(144) $\qquad |z_1|^2 + |z_2|^2 + |z_3|^2 - 3|z_0|^2 = t.$

The $U(1)$ action on the zs is

(145)
$$(z_0, z_1, z_2, z_3) \rightarrow (e^{-3i\alpha} z_0, e^{i\alpha} z_1, e^{i\alpha} z_2, e^{i\alpha} z_3).$$

Notice that $z_{1,2,3}$ describe the basis $\mathbb{P}^2$, while z_0 parameterizes the complex direction of the fibre.

Let us now give a description in terms of glued $\mathbf{C}^3$ patches. There are three patches U_i defined by $z_i \neq 0$, for $i = 1, 2, 3$, since at least one of these three co-ordinates must be non-zero in X. All of these three patches look like $\mathbf{C}^3$. For example, for $z_3 \neq 0$, we can 'solve' again for z_3 in terms of the other three unconstrained co-ordinates that then parameterize $\mathbf{C}^3$: $U_3 = (z_0, z_1, z_2)$. Similar statements hold for the other two patches. Let us now construct the corresponding degeneration graph. In the $U_3 = (z_0, z_1, z_2)$ patch we take as our Hamiltonians

(146)
$$r_\alpha = |z_1|^2 - |z_0|^2,$$
$$r_\beta = |z_2|^2 - |z_0|^2.$$

The graph of the degenerate fibres in the $r_\alpha - r_\beta$ plane is the same as in the $\mathbf{C}^3$ example, Fig. 12. The third direction in the base, r_γ is now given by the gauge invariant product $r_\gamma = \mathrm{Im}(z_0 z_1 z_2 z_3)$. The same two Hamiltonians $r_{\alpha,\beta}$

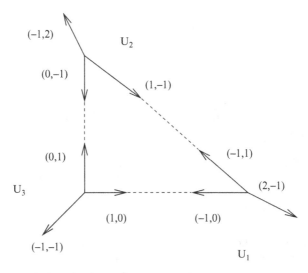

Fig. 14. The graph of $\mathcal{O}(-3) \rightarrow \mathbb{P}^2$. This manifold is built out of three $\mathbf{C}^3$ patches

generate the action in the $U_2 = (z_0, z_1, z_3)$ patch, and we use the constraint (144) to rewrite them as follows: since both z_0 and z_1 are co-ordinates of this patch r_α does not change. On the other hand, r_β must be rewritten since z_2 is not a natural co-ordinate here. We then find:

$$r_\alpha = |z_1|^2 - |z_0|^2,$$

(147)
$$r_\beta = t + 2|z_0|^2 - |z_1|^2 - |z_3|^2,$$

hence

$$e^{\alpha r_\alpha + \beta r_\beta} : (z_0, z_1, z_3) \rightarrow (e^{i(-\alpha+2\beta)} z_0, e^{i(\alpha-\beta)} z_1, e^{-i\beta} z_3).$$

We see from the above that the fibres degenerate over three lines (1) $r_\alpha + r_\beta = t$, corresponding to $z_0 = 0 = z_3$, and where a $(-1,1)$ cycle degenerates; (2) there is a line over which a $(-1,2)$ cycle degenerates where $z_1 = 0 = z_3$, $2r_\alpha + r_\beta = t$, and finally, (3) there is a line over which $r_\alpha = 0$, and a $(0,1)$-cycle degenerates. The U_1 patch is similar, and we end up with the graph for $\mathcal{O}(-3) \rightarrow \mathbb{P}^2$ shown in Fig. 14.

Example. *Lagrangian submanifolds.* In order to consider open string amplitudes in the above Calabi–Yau geometries, we have to construct Lagrangian submanifolds providing boundary conditions, as we explained in Sect. 4.4. Let us start by considering the $\mathbf{C}^3$ geometry discussed above. In this case, one can easily construct Lagrangian submanifolds following the work of Harvey and Lawson (1982). In terms of the Hamiltonians in (131), we have three types of them:

$$
\begin{array}{llll}
L_1: & r_\alpha = 0, & r_\beta = r_1, & r_\gamma \geq 0. \\
L_2: & r_\alpha = r_2, & r_\beta = 0, & r_\gamma \geq 0. \\
L_3: & r_\alpha = r_\beta = r_3, & r_\gamma \geq 0,
\end{array}
$$

(148)

where r_i, $i = 1, 2, 3$, are constants. It is not difficult to check that the above submanifolds are indeed Lagrangian (they turn out to be Special Lagrangian as well). In terms of the graph description we developed above, they correspond to points in the edges of the planar graph spanned by (r_α, r_β), and they project to semi-infinite straight lines on the basis of the fibration, $\mathbf{R}^3$, parameterized by $r_\gamma \geq 0$. Since they are located at the edges, where one of the circles of the fibration degenerates, they have the topology of $\mathbf{C} \times \mathbf{S}^1$.

It is easy to generalize the construction to other toric geometries, like the resolved conifold or local $\mathbb{P}^2$: Lagrangian submanifolds with the topology of $\mathbf{C} \times \mathbf{S}^1$ are just given by points on the edges of the planar graphs. Such Lagrangian submanifolds were first considered in the context of open topological string theory by Aganagic and Vafa (2000), and further studied by Aganagic et al. (2002).

4.3 Examples of Closed String Amplitudes

Now that we have presented some detailed constructions of Calabi–Yau threefolds, we can come back totopological string amplitudes, or equivalently to Gromov–Witten invariants. The Gromov–Witten invariants of Calabi–Yau threefolds can be computed in a variety of ways. A powerful technique that can be made mathematically rigorous is the localization technique pioneered by Kontsevich (1995). For compact Calabi–Yau manifolds, only $N_{g=0,\beta}$ have been computed in detail, but for non-compact, toric Calabi–Yau manifolds localization techniques make it possible to compute $N_{g,\beta}$ for arbitrary genus. We will now present some results for the topological string amplitudes F_g of the geometries we described above.

The resolved conifold $\mathcal{O}(-1) \oplus \mathcal{O}(-1) \to \mathbb{P}^1$ has a single Kähler parameter t corresponding to the $\mathbb{P}^1$ in the base, and its total free energy is given by

$$(149) \qquad F(g_s, t) = \sum_{d=1}^{\infty} \frac{1}{d \left(2 \sin \frac{dg_s}{2} \right)^2} Q^d,$$

where $Q = \mathrm{e}^{-t}$. We see that the only non-zero Gopakumar–Vafa invariant occurs at degree one and genus zero and is given by $n_1^0 = 1$. On the other hand, this model already has an infinite number of non-trivial $N_{g,\beta}$ invariants, which can be obtained by expanding the above expression in powers of g_s. The above expression was obtained in Gromov–Witten theory by Faber and Pandharipande (2000).

The space $\mathcal{O}(-3) \to \mathbb{P}^2$ also has one single Kähler parameter, corresponding to the hyperplane class of $\mathbb{P}^2$. By using the localization techniques of Kontsevich, adapted to the non-compact case, one finds (Chiang et al. 1999; Klemm and Zaslow 2001)

$$F_0(t) = -\frac{t^3}{18} + 3\,Q - \frac{45\,Q^2}{8} + \frac{244\,Q^3}{9} - \frac{12333\,Q^4}{64} \cdots$$

$$F_1(t) = -\frac{t}{12} + \frac{Q}{4} - \frac{3\,Q^2}{8} - \frac{23\,Q^3}{3} + \frac{3437\,Q^4}{16} \cdots$$

$$(150) \qquad F_2(t) = \frac{\chi(X)}{5720} + \frac{Q}{80} + \frac{3\,Q^3}{20} + \frac{514\,Q^4}{5} \cdots,$$

and so on. In (150), t is the Kähler class of the manifold, $Q = \mathrm{e}^{-t}$, and $\chi(X) = 2$ is the Euler characteristic of local $\mathbb{P}^2$. The first term in F_0 is proportional to the intersection number H^3 of the hyperplane class, while the first term in F_1 is proportional to the intersection number $H \cdot c_2(X)$. The first term in F_2 is the contribution of constant maps.

As we explained above, we can express the closed string amplitudes in terms of Gopakumar–Vafa invariants. Let us introduce a generating functional for integer invariants as follows:

$$(151) \qquad f(z, Q) = \sum_{g,\beta} n_\beta^g z^g Q^\beta,$$

where z is a formal parameter. For local $\mathbb{P}^2$ one finds

(152)
$$f(z, Q) = 3\,Q - 6\,Q^2 + (27 - 10\,z)\,Q^3 - (192 - 231\,z + 102\,z^2 - 15\,z^3)\,Q^4 + \mathcal{O}(Q^5).$$

It should be mentioned that there is a very powerful method to compute the amplitudes F_g, namely mirror symmetry. In the mirror symmetric computation, the F_g amplitudes are deeply related to the variation of complex structures on the Calabi–Yau manifold (Kodaira–Spencer theory) and can be computed through the holomorphic anomaly equations of Bershadsky et al. (1993, 1994). Gromov–Witten invariants of non-compact, toric Calabi–Yau threefolds have been computed with mirror symmetry by Chiang et al. (1999), Klemm and Zaslow (2001) and Katz et al. (1999).

5 The Topological Vertex

5.1 The Gopakumar–Vafa Duality

For topological string theory on the resolved conifold, the result in (149) shows that there is only one nontrivial Gopakumar–Vafa invariant. If we now take into account (111), we see that the free energies $F_g(t)$ are precisely the resummed functions (33) of Chern–Simons theory, after we identify the string coupling constant g_s with the gauge theory coupling constant as in (81), and the Kähler parameter of the resolved conifold is identified with the 't Hooft coupling

$$(153) \qquad\qquad t = \mathrm{i}g_s N = xN.$$

Based on this and other evidence, Gopakumar and Vafa (1998b) conjectured that *Chern–Simons theory on* $\mathbf{S}^3$ *is equivalent to closed topological string theory on the resolved conifold.*

From the point of view of topological string theory, this equivalence only illuminates the resolved conifold geometry, which on the other hand is easy to compute. The fundamental question is: can we use this duality to obtain information about more general Calabi–Yau threefolds? The answer is yes, and the underlying reasoning is heavily based on the idea of *geometric transitions*, which we won't explain here (see Mariño 2005, for a detailed exposition). This line of reasoning leads directly to the idea of the topological vertex.

5.2 Framing of Topological Open String Amplitudes

As we will see, the topological vertex is an open string amplitude, and in order to understand it properly we have to discuss one aspect of open string amplitudes that we have not addressed yet: the *framing ambiguity*. The framing ambiguity was discovered by Aganagic et al. (2002). They realized that when the boundary conditions are specified by non-compact Lagrangian submanifolds like the ones described in (148), the corresponding topological open string amplitudes are not unambiguously defined: they depend on a choice of an integer (more precisely, one integer for each boundary). For the Lagrangian submanifolds considered in Sect. 5, the framing ambiguity can be specified by specifying a vector $'f = (p, q)$ attached to the edge where the submanifold is located (see for example Mariño (2005) for a full justification of this). The procedure is illustrated in Fig. 15. It is useful to introduce the symplectic

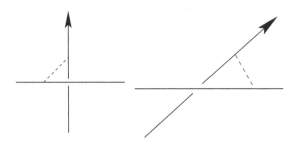

Fig. 15. Specifying a framing

product of two vectors $v = (v_1, v_2)$ and $w = (w_1, w_2)$ as

$$(154) \qquad v \wedge w = v_1 w_2 - v_2 w_1.$$

This product is invariant under $SL(2, \mathbf{Z})$ transformations. If the original Lagrangian submanifold is located at an edge v, the choice of framing has to satisfy

$$(155) \qquad f \wedge v = 1.$$

Clearly, if f satisfies (155), so does $f - nv$ for any integer n. The choice of the integer n is precisely the framing ambiguity found by Aganagic et al. (2002). In the case of the Lagrangian submanifolds of $\mathbf{C}^3$ that we constructed in Sect. 5, A particular choice of framing that will be very important in the following is shown in Fig. 16.

What is the effect of a change of framing on open topological string amplitudes? A proposal for this was made by Aganagic et al. (2002) and further studied by Mariño and Vafa (2002), based on the duality with Chern–Simons theory. As pointed out by Ooguri and Vafa (2000), vacuum expectation values of Wilson loops in Chern–Simons theory on $\mathbf{S}^3$ compute open string amplitudes. On the other hand, we explained in Sect. 2 that Wilson loop correlation functions depend on a choice of framing. This suggests that the framing ambiguity of Chern–Simons theory corresponds to the ambiguity of topological open string amplitudes that we have just described. This also leads to a very precise prescription to compute the effect of a change of framing for open string amplitudes. Let us consider for simplicity an open string amplitude involving a single Lagrangian submanifold, computed for a framing f. If we now consider the framing $f - nv$, the coefficients Z_R of the total partition function (120) change as follows

$$(156) \qquad Z_R \to (-1)^{n\ell(R)} q^{\frac{n\kappa_R}{2}} Z_R,$$

where κ_R was defined in (74), and $q = e^{ig_s}$. This is essentially the behaviour of Chern–Simons invariants under change of framing spelled out in (75). The extra sign in (156) is crucial to guarantee integrality of the resulting amplitudes, as was verified in Aganagic et al. (2002) and Mariño and Vafa (2002). If the open string amplitudes involves L boundaries, one has to specify L different framings n_α, and (156) is generalized to

$$(157) \qquad Z_{R_1 \cdots R_L} \to (-1)^{\sum_{\alpha=1}^{L} n_\alpha \ell(R_\alpha)} q^{\sum_{\alpha=1}^{L} n_\alpha \kappa_{R_\alpha}/2} Z_{R_1 \cdots R_L}.$$

5.3 Definition of the Topological Vertex

In Sect. 5 we showed that we can construct one Lagrangian submanifold in each of the vertices of the toric diagram of $\mathbf{C}^3$. Since each of these submanifolds has the topology of $\mathbf{C} \times \mathbf{S}^1$, we can consider the topological *open* string amplitude associated to this geometry. The total open string partition function will be given by

$$(158) \qquad Z(V_i) = \sum_{R_1, R_2, R_3} C_{R_1 R_2 R_3} \prod_{i=1}^{3} \operatorname{Tr}_{R_i} V_i,$$

where V_i is a matrix source associated to the i-th Lagrangian submanifold. The amplitude $C_{R_1 R_2 R_3}$ is a function of the string coupling constant g_s and, in the genus expansion, it contains information about maps from Riemann surfaces of arbitrary genera into $\mathbf{C}^3$ with boundaries on L_i. This open string amplitude is called the *topological vertex*, and it is the basic object from which, by gluing, one can obtain closed and open string amplitudes on arbitrary toric geometries. Since the vertex is an open string amplitude, it will depend on a choice of three different framings. As we explained in the previous section, this choice will be given by three different vectors f_1, f_2 and f_3. Let us see how to introduce this choice.

We saw in Sect. 5 that the $\mathbf{C}^3$ geometry can be represented by graphs involving three vectors v_i. These vectors can be obtained from the set in Fig. 12 by an $SL(2, \mathbf{Z})$ transformation, and satisfy (134). We will then introduce a topological vertex amplitude $C_{R_1 R_2 R_3}^{(v_i, f_i)}$ that depends on both a choice of three vectors v_i for the edges and a choice of three vectors f_i for the framings. Due to (155) we require

$$f_i \wedge v_i = 1.$$

We will orient the edges v_i in a clockwise way. Since wedge products are preserved by $SL(2, \mathbf{Z})$, we also have

$$(159) \qquad v_2 \wedge v_1 = v_3 \wedge v_2 = v_1 \wedge v_3 = 1.$$

However, not all of these choices give independent amplitudes. First of all, there is an underlying $\mathrm{SL}(2,\mathbf{Z})$ symmetry relating the choices: if $g \in \mathrm{SL}(2,\mathbf{Z})$, then the amplitudes are invariant under

$$(f_i, v_i) \to (g \cdot f_i, g \cdot v_i).$$

Moreover, if the topological vertex amplitude $C_{R_1 R_2 R_3}^{(v_i, f_i)}$ is known for a set of framings f_i, then it can be obtained for any set of the form $f_i - n_i v_i$, and it is given by the general rule (157)

$$(160) \qquad C_{R_1 R_2 R_3}^{(v_i, f_i - n v_i)} = (-1)^{\sum_i n_i \ell(R_i)} q^{\sum_i n_i \kappa_{R_i}/2} C_{R_1 R_2 R_3}^{(v_i, f_i)},$$

for all admissible choices of the vectors v_i. Since any two choices of framing can be related through (160), it is useful to pick a convenient set of f_i for any given choice of v_i, which we will define as the *canonical framing* of the topological vertex. This canonical framing turns out to be

$$(f_1, f_2, f_3) = (v_2, v_3, v_1).$$

Due to the $\mathrm{SL}(2,\mathbf{Z})$ symmetry and the transformation rule (160), any topological vertex amplitude can be obtained from the amplitude computed for a *fixed* choice of v_i in the canonical framing. A useful choice of the v_i is $v_1 = (-1, -1), v_2 = (0, 1), v_3 = (1, 0)$, as in Fig. 12. The vertex amplitude for the canonical choice of v_i and in the canonical framing will be simply denoted by $C_{R_1 R_2 R_3}$. Any other choice of framing will be characterized by framing vectors of the form $f_i - n_i v_i$, and the corresponding vertex amplitude will be denoted by

$$C_{R_1 R_2 R_3}^{n_1, n_2, n_3}.$$

Notice that $n_i = f_i \wedge v_{i+1}$ (where i runs mod 3).

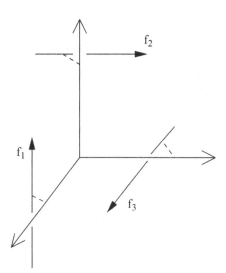

Fig. 16. The canonical choice of framing for the topological vertex

One of the most important properties of $C_{R_1 R_2 R_3}$ is its *cyclic symmetry*. To see this, notice that the SL$(2, \mathbf{Z})$ transformation $g = TS^{-1}$ takes

$$(v_i, f_i) \rightarrow (v_{i+1}, f_{i+1}),$$

where again i runs mod 3. It then follows that

$$(161) \qquad C_{R_1 R_2 R_3} = C_{R_3 R_1 R_2} = C_{R_2 R_3 R_1}.$$

Finally, it will sometimes be useful to consider the vertex in the basis of conjugacy classes $C_{\mathbf{k}^{(1)} \mathbf{k}^{(2)} \mathbf{k}^{(3)}}$, which is obtained from $C_{R_1 R_2 R_3}$ by

$$(162) \qquad C_{\mathbf{k}^{(1)} \mathbf{k}^{(2)} \mathbf{k}^{(3)}} = \sum_{R_i} \prod_{i=1}^{3} \chi_{R_i}(C(\mathbf{k}^{(i)})) C_{R_1 R_2 R_3}.$$

5.4 Gluing Rules

We saw in Sect. 5 that any non-compact toric geometry can be encoded in a planar graph that can be obtained by gluing trivalent vertices. It is then natural to expect that the string amplitudes associated to such a diagram can be computed by gluing the open topological string amplitudes associated to the trivalent vertices, in the same way that one computes amplitudes in perturbative quantum field theory by gluing vertices through propagators. This idea was suggested by Aganagic et al. (2004) and Iqbal (2002), and was developed into a complete set of rules by Aganagic et al. (2005). The gluing rules for the topological vertex turn out to be quite simple. Here we will state three rules (for a change of orientation in one edge, for the propagator, and for the matching of framings in the gluing) which make it possible to compute any closed string amplitude on toric, non-compact Calabi–Yau threefolds. They also make it possible to compute open string amplitudes for Lagrangian submanifolds on edges that go to infinity. The case of Lagrangian submanifolds on inner edges is also very easy to analyze, but we refer the reader to the paper by Aganagic, Klemm, Mariño, and Vafa (2005) for the details. A mathematical point of view on the gluing rules can be found in Diaconescu and Florea (2005) and Li et al. (2004):

1) *Orientation.* Trivalent vertices are glued along their edges, and this corresponds to gluing curves with holes along their boundaries. In order to do that, the boundaries must have opposite orientations. This change of orientation will be represented as an inversion of the edge vector, therefore in gluing the vertices we will have an outgoing edge on one side, say v_1, and an ingoing

edge on the other side, $-v_1$. It can be shown that the under this operation the topological vertex changes as

$$C_{R_1 R_2 R_3} \rightarrow (-1)^{\ell(R_1)} C_{R_1^t R_2 R_3}.$$

Of course, a similar equation follows for the other v_i.

2) *Propagator.* Since gluing the edges corresponds to gluing curves with holes along their boundaries, we must have matching of the number of holes and winding numbers along the edge. After taking into account the change of orientation discussed above, and after a simple analysis (Aganagic et al. 2005), one finds that the propagator for gluing edges with representations R_1, R_2 is given by

$$(163) \qquad\qquad (-1)^{\ell(R_1)} e^{-\ell(R_1)t} \delta_{R_1 R_2^t},$$

where t is the Kähler parameter that corresponds to the $\mathbb{P}^1$ represented by the gluing edge.

3) *Framing.* When gluing two vertices, the framings of the two edges involved in the gluing have to match. This means that, in general, we will have to change the framing of one of the vertices. Let us consider the case in which we glue together two vertices with outgoing vectors (v_i, v_j, v_k) and (v_i', v_j', v_k'), respectively, and let us assume that we glue them through the vectors v_i, $v_i' = -v_i$. We also assume that both vertices are canonically framed, so that $f_i = v_j$, $f_i' = v_j'$. In order to match the framings we have to change the framing of, say, v_i', so that the new framing is $-f_i$ (the opposite sign is again due to the change of orientation). There is an integer n_i such that $f_i' - n_i v_i' = -f_i$ (since $f_i \wedge v_i = f_i' \wedge v_i' = 1$, $f_i + f_i'$ is parallel to v_i), and it is easy to check that

$$n_i = v_j' \wedge v_j.$$

The gluing of the two vertex amplitudes is then given by

$$(164) \qquad \sum_{R_i} C_{R_j R_k R_i} e^{-\ell(R_i)t_i} (-1)^{(n_i+1)\ell(R_i)} q^{-n_i \kappa_{R_i}/2} C_{R_i^t R_j' R_k'},$$

where we have taken into account the change of orientation in the (v_i', v_j', v_k') to perform the gluing, and t_i is the Kähler parameter associated to the edge.

Given then a planar trivalent graph representing a non-compact Calabi–Yau manifold, we can compute the closed string amplitude as follows: we give a presentation of the graph in terms of vertices glued together, as we did in Sect. 5. We associate the appropriate amplitude to each trivalent vertex (labelled by representations), and we use the above gluing rules. The edges that go to infinity carry the trivial representation, and we finally sum over all possible representations along the inner edges. The resulting quantity is the total partition function $Z_{\text{closed}} = e^F$ for closed string amplitudes.

5.5 Explicit Expression for the Topological Vertex

Once we have defined the topological vertex, we need an explicit expression for it. This turns out to be a difficult problem which can however be explicitly solved. The basic idea is to use an extension of the Gopakumar–Vafa duality to open string amplitudes. As shown by Ooguri and Vafa (2000), the duality between Chern–Simons theory and the resolved conifold leads to a correspondence between *Chern–Simons invariants for knots in* $\mathbf{S}^3$ and *open topological string amplitudes with Lagrangian boundary conditions in the resolved conifold*. This idea applies in principle to Lagrangian submanifolds in the resolved conifold, but one can extend it to other contexts, and in particular to the configuration considered above involving three Lagrangian submanifolds in $\mathbf{C}^3$.

It turns out that the open topological string amplitude for the three Lagrangian submanifolds in $\mathbf{C}^3$ can be written by using only the Chern–Simons invariant of the Hopf link that we studied in Sect. 2. Let $\mathcal{W}_{R_1 R_2}$ is the Hopf link invariant defined in (54) and evaluated in (62). We now consider the limit

$$(165) \qquad W_{R_1 R_2} = \lim_{t \to \infty} \mathrm{e}^{-\frac{\ell(R_1)+\ell(R_2)}{2} t} \mathcal{W}_{R_1 R_2}.$$

This limit exists, since $\mathcal{W}_{R_1 R_2}$ is of the form $\lambda^{\frac{\ell(R_1)+\ell(R_2)}{2}} W_{R_1 R_2} + \mathcal{O}(\mathrm{e}^{-t})$ (recall that $\lambda = \mathrm{e}^t$). The quantity $W_{R_1 R_2}$, which is the 'leading' coefficient of the Hopf link invariant (54), is the building block of the topological vertex amplitude. It is a rational function of $q^{\pm \frac{1}{2}}$, therefore it only depends on the string coupling constant. We will also denote $W_R = W_{R0}$. The limit (165) was first considered by Aganagic et al. (2004). The final expression for the vertex, in the canonical framing defined above, is

$$(166) \qquad C_{R_1 R_2 R_3} = q^{\frac{\kappa_{R_2}+\kappa_{R_3}}{2}} \sum_{Q_1, Q_3, Q} N_{QQ_1}^{R_1} N_{QQ_3}^{R_3^t} \frac{W_{R_2^t Q_1} W_{R_2 Q_3}}{W_{R_2}},$$

where $N_{R_1 R_2}^R$ is the Littlewood–Richardson coefficient which gives the multiplicity of R in the tensor product $R_1 \otimes R_2$.

Let us now give some more explicit formulae for the vertex. The basic ingredient in (166) is the quantity $W_{R_1 R_2}$ defined in (165). Using (62) it is possible to give an explicit expression for $W_{R_1 R_2}$ that is useful in computations. It is easy to see that the leading coefficient of λ in (62) is obtained by taking the leading coefficient of λ in $\dim_q R_2$ and the λ-independent piece in (65). The generating function of elementary symmetric polynomials (63) then becomes

$$(167) \qquad S(t) \prod_{j=1}^{c_R} \frac{1 + q^{l_j^R - j} t}{1 + q^{-j} t},$$

where

(168) $$S(t) = \prod_{j=1}^{\infty}(1 + q^{-j}t) = 1 + \sum_{r=1}^{\infty} \frac{q^{-\frac{r(r+1)}{2}}t^r}{\prod_{m=1}^{r}[m]}.$$

Notice that (167) is the generating function of elementary symmetric polynomials for an infinite number of variables given by $x_j = q^{l_j^{R_1}-j}$, $j = 1, 2, \cdots$. One then deduces that the $\lambda \to \infty$ limit of $q^{\ell(R_1)/2}s_{R_1}(x_i = q^{l_i^{R_2}-i})$ is given by the Schur polynomial

$$s_{R_1}(x_i = q^{l_i^{R_2}-i+\frac{1}{2}}),$$

which now involves an *infinite* number of variables x_i. This finally leads to the following expression for $W_{R_1 R_2}$:

(169) $$W_{R_1 R_2}(q) = s_{R_1}(x_i = q^{l_i^{R_2}-i+\frac{1}{2}})s_{R_2}(x_i = q^{-i+\frac{1}{2}}).$$

We will also write this as

(170) $$W_{R_1 R_2}(q) = s_{R_1}(q^{\rho+l^{R_2}})s_{R_2}(q^{\rho}),$$

where the arguments of the Schur functions indicate the above values for the polynomial variables x_i. Using (170) one can write (166) in terms of skew Schur polynomials (Okounkov et al. 2003):

(171) $$C_{R_1 R_2 R_3} = q^{\frac{1}{2}(\kappa_{R_2}+\kappa_{R_3})}s_{R_2^t}(q^{\rho})\sum_Q s_{R_1/Q}(q^{\ell(R_2^t)+\rho})s_{R_3^t/Q}(q^{\ell(R_2)+\rho}).$$

5.6 Applications

We will now present some examples of computation of topological string amplitudes by using the topological vertex.

Example. *Resolved conifold.* The toric diagram for the resolved conifold geometry is depicted in Fig. 13. Our rules give immediately:

(172) $$Z_{\mathbb{P}^1} = \sum_R C_{00R^t}(-1)^{\ell(R)}e^{-\ell(R)t}C_{R00}.$$

Since $C_{R00} = W_R = s_R(x_i = q^{-i+\frac{1}{2}})$, we can use well–known formulae for Schur polynomials to obtain

(173) $$Z_{\mathbb{P}^1} = \exp\left\{-\sum_{d=1}^{\infty} \frac{e^{-dt}}{d(q^{\frac{d}{2}} - q^{-\frac{d}{2}})^2}\right\},$$

in agreement with the known result (149).

Example. *Local* $\mathbb{P}^2$. The toric diagram is depicted in Fig. 14. Using again the rules explained above, we find the total partition function

$$(174) \quad Z_{\mathbb{P}^2} = \sum_{R_1,R_2,R_3} (-1)^{\sum_i \ell(R_i)} e^{-\sum_i \ell(R_i)t} q^{-\sum_i \kappa_{R_i}} C_{0R_2^t R_3} C_{0R_1^t R_2} C_{0R_3^t R_1},$$

where t is the Kähler parameter corresponding to the hyperplane class in $\mathbb{P}^2$. Using that $C_{0R_2 R_3^t} = W_{R_2 R_3} q^{-\kappa_{R_3}/2}$ one recovers the expression for $Z_{\mathbb{P}^2}$ first obtained by Aganagic et al. (2004). Notice that the free energy has the structure

$$(175) \qquad F_{\mathbb{P}^2} = \log \left\{ 1 + \sum_{\ell=1}^{\infty} a_\ell(q) e^{-\ell t} \right\} = \sum_{\ell=1}^{\infty} a_\ell^{(c)}(q) e^{-\ell t}.$$

The coefficients $a_\ell(q)$, $a_\ell^{(c)}(q)$ can be easily obtained in terms of $W_{R_1 R_2}$. One finds, for example,

$$a_1^{(c)}(q) = a_1(q) = -\frac{3}{(q^{\frac{1}{2}} - q^{-\frac{1}{2}})^2},$$

$$(176) \qquad a_2^{(c)}(q) = \frac{6}{(q^{\frac{1}{2}} - q^{-\frac{1}{2}})^2} + \frac{1}{2} a_1(q^2).$$

If we compare to (110) and take into account the effects of multicovering, we find the following values for the Gopakumar–Vafa invariants of $\mathcal{O}(-3) \to \mathbb{P}^2$:

$$\begin{matrix} n_1^0 = 3, & n_1^g = 0 \text{ for } g > 0, \\ (177) \qquad n_2^0 = -6, & n_2^g = 0 \text{ for } g > 0, \end{matrix}$$

in agreement with the results listed in (152). In fact, one can go much further with this method and compute the Gopakumar–Vafa invariants to high degree. We again see that the use of exact results in Chern–Simons theory leads to the topological string amplitudes to *all genera*. A complete listing of the Gopakumar–Vafa invariants up to degree 12 can be found in Aganagic et al. (2004). The partition function (174) can also be computed in Gromov–Witten theory by using localization techniques, and one finds indeed the same result (Zhou 2003).

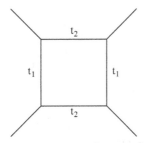

Fig. 17. The toric diagram of local $\mathbb{P}^1 \times \mathbb{P}^1$

Example. *Local* $\mathbb{P}^1 \times \mathbb{P}^1$. The local $\mathbb{P}^1 \times \mathbb{P}^1$ geometry is the non-compact Calabi–Yau manifold given by the four-manifold $\mathbb{P}^1 \times \mathbb{P}^1$ together with its anti-canonical bundle. It also admits a symplectic quotient description of the form (137), this time with $N = 2$ and two Kähler parameters t_1, t_2. The charges $Q_{1,2}^j$, $j = 1, \cdots, 5$ can be grouped into two vectors

$$Q_1 = (-2, 1, 1, 0, 0),$$
(178)
$$Q_2 = (-2, 0, 0, 1, 1).$$

The toric diagram for this geometry can be easily worked out from this description, and it is represented in Fig. 17. Using the gluing rules we find the closed string partition function

$$Z_{\mathbb{P}^1 \times \mathbb{P}^1} = \sum_{R_i} e^{-(\ell(R_1)+\ell(R_3))t_1 - (\ell(R_2)+\ell(R_4))t_2} q^{\sum_i \kappa_{R_i}/2}$$
(179)
$$\times C_{0 R_4 R_1^t} C_{0 R_1 R_2^t} C_{0 R_2 R_3^t} C_{0 R_3 R_4^t}.$$

This amplitude can be written as

$$Z_{\mathbb{P}^1 \times \mathbb{P}^1} = \sum_{R_i} e^{-(\ell(R_1)+\ell(R_3))t_1 - (\ell(R_2)+\ell(R_4))t_2}$$
(180)
$$\times W_{R_4 R_1} W_{R_1 R_2} W_{R_2 R_3} W_{R_3 R_4}.$$

This is the expression first obtained by Aganagic et al. (2004). It has been shown to agree with Gromov–Witten theory by Zhou (2003).

Example. *The closed topological vertex.* Consider the Calabi–Yau geometry whose toric diagram is depicted in Fig. 18. It contains three $\mathbb{P}^1$ touching at a single point. The local Gromov–Witten theory of this geometry was studied by Bryan and Karp (2005), who called it the closed topological vertex, and also by Karp et al. (2005). The vertex rules give the following expression for the total partition function:

(181)
$$Z(t_1, t_2, t_3) = \sum_{R_1, R_2, R_3} C_{R_1 R_2 R_3} W_{R_1^t} W_{R_2^t} W_{R_3^t} (-1)^{\ell(R_1)+\ell(R_2)+\ell(R_3)} e^{-\sum_{i=1}^3 \ell(R_i) t_i}.$$

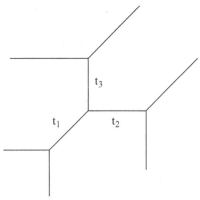

Fig. 18. The toric diagram of the closed topological vertex

It turns out that this can be evaluated in closed form (Karp et al. 2005)

$$
\begin{aligned}
Z(t_1, t_2, t_3) = \exp\Bigg(&-\sum_{d=1}^{\infty} \frac{1}{d(q^{\frac{d}{2}} - q^{-\frac{d}{2}})} \Big(e^{-dt_1} + e^{-dt_2} + e^{-dt_3} \\
&- e^{-d(t_1+t_2)} - e^{-d(t_1+t_3)} - e^{-d(t_2+t_3)} + e^{-d(t_1+t_2+t_3)}\Big)\Bigg),
\end{aligned}
$$

(182)

in agreement with the algebro-geometric results of Bryan and Karp (2005). Notice from the above expression that there is only a finite number of non-vanishing Gopakumar–Vafa invariants for the above geometry.

A Symmetric Polynomials

In this brief Appendix we summarize some useful ingredients of the elementary theory of symmetric functions. A standard reference is Macdonald (1995).

Let $x_1, \cdots, x_N$ denote a set of N variables. The *elementary symmetric polynomials* in these variables, $e_m(x)$, are defined as:

$$
(183) \qquad e_m(x) = \sum_{i_1 < \cdots < i_m} x_{i_1} \cdots x_{i_m}.
$$

The generating function of these polynomials is given by

$$
(184) \qquad E(t) = \sum_{m \geq 0} e_m(x) t^m = \prod_{i=1}^{N} (1 + x_i t).
$$

The *complete symmetric function* h_m can be defined in terms of its generating function

$$
(185) \qquad H(t) = \sum_{m \geq 0} h_m t^m = \prod_{i=1}^{N} (1 - x_i t)^{-1},
$$

and one has

$$
(186) \qquad E(t) H(-t) = 1.
$$

The products of elementary symmetric polynomials and of complete symmetric functions provide two different basis for the symmetric functions of N variables.

Another basis is given by the *Schur polynomials*, $s_R(x)$, which are labelled by representations R. We will always express these representations in terms of Young tableaux, so R is given by a partition $(l_1, l_2, \cdots, l_{c_R})$, where l_i is the number of boxes of the i-th row of the tableau, and we have $l_1 \geq l_2 \geq \cdots \geq l_{c_R}$.

The total number of boxes of a tableau will be denoted by $\ell(R) = \sum_i l_i$. The Schur polynomials are defined as quotients of determinants,

$$(187) \qquad s_R(x) = \frac{\det x_j^{l_i+N-i}}{\det x_j^{N-i}}.$$

They can be written in terms of the symmetric polynomials $e_i(x_1, \cdots, x_N)$, $i \geq 1$, as follows:

$$(188) \qquad s_R = \det M_R,$$

where

$$M_R^{ij} = (e_{l_i^t+j-i}).$$

M_R is an $r \times r$ matrix, with $r = c_{R^t}$, and R^t denotes the transposed Young tableau with row lengths l_i^t. To evaluate s_R we put $e_0 = 1$, $e_k = 0$ for $k < 0$. The expression (188) is known as the Jacobi–Trudi identity.

A third set of symmetric functions is given by the *Newton polynomials* $P_{\mathbf{k}}(x)$. These are labelled by vectors $\mathbf{k} = (k_1, k_2, \cdots)$, where the k_j are non-negative integers, and are defined as

$$(189) \qquad P_{\mathbf{k}}(x) = \prod_{j}^{p} P_j^{k_j}(x),$$

where

$$(190) \qquad P_j(x) = \sum_{i=1}^{N} x_i^j,$$

are power sums. The Newton polynomials are homogeneous of degree $\ell = \sum_j jk_j$ and give a basis for the symmetric functions in $x_1, \cdots, x_N$ with rational coefficients. They are related to the Schur polynomials through the Frobenius formula

$$(191) \qquad P_{\mathbf{k}}(x) = \sum_{R} \chi_R(C(\mathbf{k})) s_R(x),$$

where the sum is over all tableaux such that $\ell(R) = \ell$.

References

1. Aganagic, M., Klemm, A., Mariño, M., and Vafa, C. (2005). The topological vertex. *Commun. Math. Phys.* **254**, 425–478; eprint hep-th/0305132.
2. Aganagic, M., Klemm, A., and Vafa, C. (2002). Disk instantons, mirror symmetry and the duality web. Z. Naturforsch. **A 57**, 1–28; eprint hep-th/0105045.

3. Aganagic, M., Mariño, M., and Vafa, C. (2004). All loop topological string amplitudes from Chern–Simons theory. *Commun. Math. Phys.* **247**, 467–512; eprint hep-th/0206164.

4. Aganagic, M. and Vafa, C. (2000). Mirror symmetry, D-branes and counting holomorphic discs. Eprint hep-th/0012041.

5. Aspinwall, P. S., and Morrison, D. R. (1993). Topological field theory and rational curves. *Commun. Math. Phys.* **151**, 245–62; eprint hep-th/9110048.

6. Atiyah, M. (1990). On framings of three-manifolds. *Topology* **29**, 1–7.

7. Axelrod, S. and Singer, I. (1992). Chern–Simons perturbation theory. In Catto, Sultan (ed.) *et al.*, *Differential geometric methods in theoretical physics*, World Scientific, Singapore, pp. 3–45; eprint hep-th/9110056.

8. Bar-Natan, D. (1995). On the Vassiliev knot invariants. *Topology* **34**, 423–72.

9. Bershadsky, M., Cecotti, S., Ooguri, H., and Vafa, C. (1993). Holomorphic anomalies in topological field theories. *Nucl. Phys. B* **405**, 279–304; eprint hep-th/9302103.

10. Bershadsky, M., Cecotti, S., Ooguri, H., and Vafa, C. (1994). Kodaira–Spencer theory of gravity and exact results for quantum string amplitudes. *Commun. Math. Phys.* **165**, 311–428; eprint hep-th/9309140.

11. Bryan, J. and Karp, D. (2005). The closed topological vertex via the Cremona transform. *J. Algebraic Geom.* **14**, 529–42; eprint math.AG/0311208.

12. Bryan, J. and Pandharipande, R. (2001). BPS states of curves in Calabi–Yau 3-folds. *Geom. Topol.* **5**, 287–318; eprint math.AG/0009025.

13. Camperi, M., Levstein, F., and Zemba, G. (1990). The large N limit of Chern–Simons theory. *Phys. Lett. B* **247**, 549–54.

14. Candelas, P., De La Ossa, X. C., Green, P., and Parkes, L. (1991). A pair of Calabi–Yau manifolds as an exactly soluble superconformal theory. *Nucl. Phys. B* **359**, 21–74.

15. Chiang, T. M., Klemm, A., Yau, S. T., and Zaslow, E. (1999). Local mirror symmetry: Calculations and interpretations. *Adv. Theor. Math. Phys.* **3**, 495–565; eprint hep-th/9903053.

16. Correale, R. and Guadagnini, E. (1994). Large N Chern–Simons field theory. *Phys. Lett. B* **337**, 80–5.

17. Cox, D. and Katz, S. 1999. *Mirror symmetry and algebraic geometry*, American Mathematical Society, Providence.

18. Cvitanovic, P. (1976). Group theory for Feynman diagrams in nonabelian gauge theories. *Phys. Rev. D* **14**, 1536–53.

19. Cvitanovic, P. (2004). *Group theory. Birdtracks, Lie's, and exceptional groups.* In http://www.cns.gatech.edu/GroupTheory.

20. Diaconescu, D. E. and Florea, B. (2005). Localization and gluing of topological amplitudes. *Commun. Math. Phys.* **257**, 119–49; eprint hep-th/0309143.

21. Di Francesco, P., Mathieu, P., and Sénéchal, D. (1997). *Conformal field theory*, Springer-Verlag, New York.

22. Dijkgraaf, R. (1995). Perturbative topological field theory. In *String theory, gauge theory and quantum gravity '93*, 189–227, World Scientific.

23. Elitzur, S., Moore, G., Schwimmer, A., and Seiberg, N. (1989). Remarks on the canonical quantization of the Chern–Simons-Witten theory. *Nucl. Phys. B* **326**, 108–34.

24. Faber, C. and Pandharipande, R. (2000). Hodge integrals and Gromov–Witten theory. *Invent. Math.* **139**, 173–99; eprint math.AG/9810173.

25. Freed, D. S. and Gompf, R. E. (1991). Computer calculation of Witten's three-manifold invariant. *Commun. Math. Phys.* **141**, 79–117.

26. Freyd, P., Yetter, D., Hoste, J., Lickorish, W.B.R., Millett, K., and Ocneanu, A.A. (1985). A new polynomial invariant of knots and links. *Bull. Am. Math. Soc. (N.S.)* **12**, 239–46.

27. Fulton, W. and Harris, J. (1991). *Representation theory. A first course.* Springer-Verlag, New York.

28. Gopakumar, R., and Vafa, C. (1998a). M-theory and topological strings, I. Eprint hep-th/9809187.

29. Gopakumar, R. and Vafa, C. (1998b). M-theory and topological strings, II. Eprint hep-th/9812127.

30. Gopakumar, R. and Vafa, C. (1999). On the gauge theory/geometry correspondence. *Adv. Theor. Math. Phys.* **3**, 1415–43; eprint hep-th/9811131.

31. Graber, T. and Zaslow, E. (2002). Open-string Gromov-Witten invariants: Calculations and a mirror "theorem". Contemp. Math. **310**, 107–121; eprint hep-th/0109075.

32. Griffiths, P. and Harris, J. (1977). *Principles of algebraic geometry*, John Wiley and Sons, New York.

33. Guadagnini, E., Martellini, M., and Mintchev, M. (1990). Wilson lines in Chern–Simons theory and link invariants. *Nucl. Phys. B* **330**, 575–607.

34. Harris, J. and Morrison, I. (1998). *Moduli of curves.* Springer-Verlag, New York.

35. Harvey, R. and Lawson, H.B. (1982). Calibrated geometries. *Acta Math.* **148**, 47–157.

36. Hori, K., Katz, S., Klemm, A., Pandharipande, R., Thomas, R., Vafa, C., Vakil, R., and Zaslow, E. (2003). *Mirror symmetry.* American Mathematical Society, Providence.

37. Iqbal, A. (2002). All genus topological string amplitudes and 5-brane webs as Feynman diagrams. Eprint hep-th/0207114.

38. Jeffrey, L. C. (1992). Chern–Simons-Witten invariants of lens spaces and torus bundles, and the semi-classical approximation. *Commun. Math. Phys.* **147**, 563–604.

39. Karp, D., Liu, C.-C. M., and Mariño, M. (2005). The local Gromov–Witten invariants of configurations of rational curves. Eprint math.AG/0506488.

40. Katz, S., Klemm, A., and Vafa, C. (1999). M-theory, topological strings and spinning black holes. *Adv. Theor. Math. Phys.* **3**, 1445–537; eprint hep-th/9910181.

41. Katz, S., and Liu, C.-C. (2002). Enumerative geometry of stable maps with Lagrangian boundary conditions and multiple covers of the disc. *Adv. Theor. Math. Phys.* **5**, 1–49; eprint math.ag/0103074.

42. Klemm, A. and Zaslow, E. (2001). Local mirror symmetry at higher genus. In *Winter School on Mirror Symmetry, Vector bundles and Lagrangian Submanifolds*, American Mathematical Society, Providence, p. 183–207; eprint hep-th/9906046.

43. Kontsevich, M. (1995). Enumeration of rational curves via torus actions. *Prog. Math.* **129**, 335–68; eprint hep-th/9405035.

44. Labastida, J.M.F. (1999). Chern–Simons theory: ten years after. In *Trends in Theoretical Physics II*, H. Falomir, R. E. Gamboa and F. A. Schaposnik, eds., AIP Conference Proceedings, Vol. 484, p. 1–40; eprint hep-th/9905057.

45. Labastida, J. M. F. and Ramallo, A. V. (1989). Operator formalism for Chern–Simons theories. *Phys. Lett. B* **227**, 92–102.

46. Le, T. T. Q., Murakami, J., and Ohtsuki, T. (1998). On a universal perturbative invariant of 3-manifolds. *Topology* **37**, 539–74.
47. Li, J., Liu, C.-C. M., Liu, K., and Zhou, J. (2004). A mathematical theory of the topological vertex. Eprint math.AG/0408426.
48. Li, J. and Song, Y. S. (2002). Open string instantons and relative stable morphisms. *Adv. Theor. Math. Phys.* **5**, 67–91; eprint hep-th/0103100.
49. Lickorish, W. B. R. (1998). *An introduction to knot theory*, Springer-Verlag, New York.
50. Macdonald, I.G. (1995). *Symmetric functions and Hall polynomials*, 2nd edition, Oxford University Press, Oxford.
51. Mariño, M. (2005). *Chern–Simons theory, matrix models, and topological strings*. Oxford University Press.
52. Mariño, M. and Vafa, C. (2002). Framed knots at large N. *Contemp. Math.* **310**, 185–204; eprint hep-th/0108064.
53. Maulik, D., Nekrasov, N., Okounkov, A., and Pandharipande, R. (2003). Gromov–Witten theory and Donaldson–Thomas theory. Eprint math.AG/0312059.
54. Mayr, P. (2002). Summing up open string instantons and $\mathcal{N} = 1$ string amplitudes. Eprint hep-th/0203237.
55. Morton, H.R. and Lukac, S. G. (2003). The HOMFLY polynomial of the decorated Hopf link. *J. Knot Theory Ramifications* **12**, 395–416; eprint math.GT/0108011.
56. Ohtsuki, T. (2002). *Quantum invariants*. World Scientific, Singapore.
57. Okounkov, A., Reshetikhin, N., and Vafa, C. (2003). Quantum Calabi–Yau and classical crystals. Eprint hep-th/0309208.
58. Ooguri, H. and Vafa, C. (2000). Knot invariants and topological strings. *Nucl. Phys. B* **577**, 419–38; eprint hep-th/9912123.
59. Ooguri, H. and Vafa, C. (2002). Worldsheet derivation of a large N duality. *Nucl. Phys. B* **641**, 3–34; eprint hep-th/0205297.
60. Periwal, V. (1993). Topological closed string interpretation of Chern–Simons theory. *Phys. Rev. Lett.* **71**, 1295–8; eprint hep-th/9305115.
61. Polyakov, A. M. (1988). Fermi-Bose transmutations induced by gauge fields. *Mod. Phys. Lett. A* **3**, 325–8.
62. Prasolov, V. V. and Sossinsky, A. B. (1997). *Knots, links, braids and 3-manifolds*, American Mathematical Society, Providence.
63. Rozansky, L. (1995). A large k asymptotics of Witten's invariant of Seifert manifolds. *Commun. Math. Phys.* **171**, 279–322; eprint hep-th/9303099.
64. Rozansky, L. and Witten, E. (1997). Hyper-Kähler geometry and invariants of three-manifolds. *Selecta Math.* **3**, 401–58; eprint hep-th/9612216.
65. Sawon, J. (2004). Rozansky-Witten invariants of hyperKähler manifolds. Eprint math.DG/0404360.
66. Schwarz, A. (1987). New topological invariants arising in the theory of quantized fields. Baku International Topological Conference, Abstracts (Part 2) Baku.
67. 't Hooft, G. (1974). A planar diagram theory for strong interactions. *Nucl. Phys. B* **72** 461–73.
68. Witten, E. (1989). Quantum field theory and the Jones polynomial. *Commun. Math. Phys.* **121**, 351–99.
69. Witten, E. (1993). Phases of $\mathcal{N} = 2$ theories in two dimensions. *Nucl. Phys. B* **403**, 159-222; eprint hep-th/9301042.

70. Witten, E. (1995). Chern–Simons gauge theory as a string theory. *Prog. Math.* **133**, 637–78; eprint hep-th/9207094.
71. Zhou, J. (2003). Localizations on moduli spaces and free field realizations of Feynman rules. Eprint math.AG/0310283.

Floer Cohomology with Gerbes

M. Thaddeus*

Department of Mathematics, Columbia University
2990 Broadway, New York, NY 10027, USA
thaddeus@math.columbia.edu

This is a written account of expository lectures delivered at the summer school on "Enumerative invariants in algebraic geometry and string theory" of the Centro Internazionale Matematico Estivo, held in Cetraro in June 2005. However, it differs considerably from the lectures as they were actually given. Three of the lectures in the series were devoted to the recent work of Donaldson–Thomas, Maulik–Nekrasov–Okounkov–Pandharipande, and Nakajima–Yoshioka. Since this is well documented in the literature, it seemed needless to write it up again. Instead, what follows is a greatly expanded version of the other lectures, which were a little more speculative and the least strictly confined to algebraic geometry. However, they should interest algebraic geometers who have been contemplating *orbifold cohomology* and its close relative, the so-called *Fantechi–Göttsche ring*, which are discussed in the final portion of these notes.

Indeed, we intend to argue that orbifold cohomology is essentially the same as a symplectic cohomology theory, namely *Floer cohomology*. More specifically, the quantum product structures on Floer cohomology and on the Fantechi–Göttsche ring should coincide. None of this should come as a surprise, since orbifold cohomology arose chiefly from the work of Chen–Ruan in the symplectic setting, and since the differentials in both theories involve the counting of holomorphic curves. Nevertheless, the links between the two theories are worth spelling out.

To illustrate this theme further, we will explain how both the Floer and orbifold theories can be enriched by introducing a *flat* U(1)-*gerbe*. Such a gerbe on a manifold (or orbifold) induces flat line bundles on its *loop space* and on its *inertia stack*, leading to Floer and orbifold cohomology theories with local coefficients. We will again argue that these two theories correspond. To explain all of this properly, an extended digression on the basic definitions and properties of gerbes is needed; it comprises the second of the three lectures.

---

* Supported by NSF grant DMS–0401128.

K. Behrend, M. Manetti (eds.), *Enumerative Invariants in Algebraic Geometry and String Theory*. Lecture Notes in Mathematics 1947,
© Springer-Verlag Berlin Heidelberg 2008

The plan of these notes is simple: the first lecture is a review of Floer cohomology; the second is a review of gerbes, as promised a moment ago; and the third introduces orbifold cohomology and its relatives, discusses how to add a gerbe, and interprets these constructions in terms of Floer theory. We conclude with some notes on the literature.

Since these are lecture notes, no attempt has been made to include rigorous proofs. But many aspects of Floer cohomology, especially its product structures, are not well documented in the literature either, so the reader is cautioned to take what is said about Floer cohomology with a grain of salt. The same goes for the proposed identification between the quantum product structures. It is mildly speculative but presumably should not be impossible to prove by following what has been done for the case of the identity map. Anyhow, for the moment we content ourselves with a genial narrative of a heuristic nature, making no great demands upon the reader. It presents many more definitions than theorems, but it aspires to provide a framework in which theorems may be built.

In the third lecture, I assume some familiarity with the basic definitions and properties of quantum cohomology, as given for example in the Clay Institute volume *Mirror Symmetry* (see the notes on the literature for a reference).

Acknowledgements: I thank Kai Behrend, Barbara Fantechi, and Marco Manetti for their kind invitation to speak in Cetraro, and for their patience in awaiting these notes. I am also grateful to Jim Bryan and the University of British Columbia for their hospitality while the notes were written, and to Dan Abramovich, Behrang Noohi, and Hsian-Hua Tseng for very helpful conversations and advice.

1 Floer Cohomology

This is an optimist's account of the Floer cohomology of symplectic manifolds: its origins, its construction, the main theorems, and the algebraic structures into which it naturally fits. Let me emphasize that, as an optimist's account, it presents Floer cohomology as we would like it to be, not necessarily as it is. For example, Floer proved the Arnold conjecture only in the presence of some ugly technical hypotheses, which later mathematicians have labored tirelessly to eradicate. The present account pretends that they never existed.

Floer introduced his cohomology (in fact he used homology, but never mind) to prove the Arnold conjecture on the number of fixed points of an exact Hamiltonian flow. Like so much of symplectic geometry, this problem is rooted in classical mechanics.

1.1 Newton's Second Law

Suppose a particle is moving in a time-dependent force field $\mathbf{F}(t, \mathbf{q})$. Here we regard $F : \mathbf{R} \times \mathbf{R}^3 \to \mathbf{R}^3$ as a time-dependent vector field. Newton's second law says that the trajectory $q(t)$ satisfies $F = ma$, or, taking $m = 1$ for simplicity,

$$F(t, q(t)) = \frac{d^2}{dt^2}\, q(t).$$

This is a second-order differential equation for $q(t)$. It can be easier to solve, and perhaps even visualize, such equations by the standard trick of introducing a triple of extra variables $\mathbf{p}$ and regarding the above as a first-order equation for $(p, q) \in \mathbf{R}^6$:

$$\begin{cases} F(t, q(t)) = \frac{d}{dt}\, p(t) \\ \qquad p(t) = \frac{d}{dt}\, q(t). \end{cases}$$

The solutions are flows along the time-dependent vector field on $\mathbf{R}^6$ whose value at (p, q) is $(F(t, q), p)$.

Notice that q describes the particle's position, and p describes its velocity or momentum. The space $\mathbf{R}^6$ of all (p, q) can therefore be regarded as the space of all initial conditions for the particle.

1.2 The Hamiltonian Formalism

Hamiltonian mechanics takes off from here. The idea is to cast the construction above in terms of the symplectic form on $\mathbf{R}^6 = T^*\mathbf{R}^3$, and generalize it to an arbitrary symplectic manifold.

So let M be a symplectic manifold with symplectic form ω: call it the *phase space*. Let $H : \mathbf{R} \times M \to \mathbf{R}$ be any time-dependent smooth function on M: call it the *Hamiltonian*. The symplectic form induces an isomorphism $T^*M \cong TM$; use this to make the exterior derivative $dH \in \Gamma(T^*M)$ into a vector field $V_H \in \Gamma(TM)$. Since H can depend on the time $t \in \mathbf{R}$, V_H is really a time-dependent vector field $V_H(t)$.

The *exact Hamiltonian flow* of H is the 1-parameter family of symplecto-morphisms of M

$$\Phi : \mathbf{R} \times M \to M$$

such that $d\Phi/dt = V_H(t)$ and $\Phi(0, x) = x$. Its existence and uniqueness are guaranteed by the standard theory of ODEs (at least for small t, or for M compact).

What makes this formalism so great is that it correctly describes the actual time evolution of a mechanical system when (1) the phase space M is the space of initial conditions of our system (i.e. possible positions and momenta) and (2) the Hamiltonian is the total energy, potential plus kinetic. The phase space will have a canonical symplectic form. Typically, it is of the form $M = T^*Q$ where Q parametrizes the possible configurations of the system. So one can choose (at least locally) position variables q_i and momentum variables p_i, the symplectic form is $\sum dp_i \wedge dq_i$, and kinetic energy (being essentially $\frac{1}{2}mv^2$) is some quadratic function of the momenta.

For example, from a time-independent quadratic potential on $\mathbf{R}$, you should get the simple harmonic oscillator. Or for a force $F = \nabla\psi$ on $\mathbf{R}^3$ with time-dependent potential $\psi(q_1, q_2, q_3)$, you can recover the first-order equation of the previous section by taking $H = \psi(q_1, q_2, q_3) + \frac{1}{2}(p_1^2 + p_2^2 + p_3^2)$. Or, of course, you could look at n particles moving in $\mathbf{R}^3$; then the phase space will be $\mathbf{R}^{6n}$.

Parenthetically, let's clear up a confusing detail: why is the phase space M typically identified with a cotangent bundle T^*Q and not a tangent bundle TQ? That is, why should momenta be considered cotangent vectors rather than tangent vectors? Of course, in physics we typically have a metric inducing an isomorphism between the two. Still, we may have muddied the waters somewhat by setting the mass equal to 1. The point is that momentum is a vector-valued quantity with units of $\mathrm{g\,cm\,sec}^{-1}$; it should be regarded as pairing with velocity, a vector with units of $\mathrm{cm\,sec}^{-1}$, to give energy, a scalar with units of $\mathrm{g\,cm}^2\,\mathrm{sec}^{-2}$.

1.3 The Arnold Conjecture

Arnold was interested in applications of Hamiltonian mechanics to real-life many-body problems, such as the long-term stability of the solar system. Then one is of course particularly interested in points in phase spaces that flow back to themselves, that is, $\Phi(t, x) = x$ for some $t > 0$, say $t = 1$. From now on let's write $\phi_t(x) = \Phi(t, x)$, so that $\phi_t : M \to M$ is a symplectomorphism.

Phase spaces in problems of physical interest are almost always noncompact, but Arnold realized that stronger statements might hold in the compact case. He conjectured the following:

If M is compact and ϕ_1 as above has nondegenerate fixed points, then the number of those fixed points is at least the sum of the Betti numbers of M.

Nondegeneracy of a fixed point x here means that $d\phi_1(x) - \mathrm{id}$ is nonsingular. A more general version of the Arnold conjecture, which we omit, deals with the degenerate case.

In this situation the Lefschetz fixed-point formula implies that the number of fixed points is at least the Euler characteristic, that is, the alternating sum of the Betti numbers. Hence the Arnold conjecture gives a stronger lower bound in the exact Hamiltonian case, as long as some odd Betti number is nonzero. On the other hand, if we replace the exact form dH used to define an exact Hamiltonian flow by a general closed form (you might call this a *closed Hamiltonian flow*, but you can easily check that all 1-parameter families of symplectomorphisms starting at the identity are of this kind), then the Arnold conjecture is false. Just consider the linear flow on a torus.

1.4 Floer's Proof

Floer defines cohomology groups $HF^*(\phi)$ associated to any symplectomorphism, and shows:

(1) That $HF^*(\mathrm{id}) = H^*(M)$.

(2) That for ϕ with nondegenerate fixed points, $HF^*(\phi)$ can be calculated from a complex whose chains are formal linear combinations of fixed points.

(3) That $HF^*(\phi)$ is, in a suitable sense, invariant under composition with an exact Hamiltonian flow.

The Arnold conjecture is an immediate consequence, as the dimension of a chain complex must be at least the dimension of its cohomology.

The chain complex leading to this cohomology theory is an infinite-dimensional analogue of the Morse complex, so let's pause first to review the salient points about that.

1.5 Morse Theory

Let X be a compact oriented manifold of finite dimension n. A *Morse function* $f : X \to \mathbf{R}$ is a smooth function with isolated critical points, at each of which the Hessian is nondegenerate. The *Hessian* is the matrix of second partials, but never mind: just recall instead that, according to the Morse lemma, this nondegeneracy is equivalent to the existence of local coordinates $x_1, \ldots, x_n$ in which

$$f(x_1, \ldots, x_n) = -x_1^2 - x_2^2 - \cdots - x_m^2 + x_{m+1}^2 + \cdots + x_n^2.$$

The number of negative terms is called the *Morse index*.

Let C be the set of formal linear combinations of the critical points x_i (with, say, complex coefficients). This is a finite-dimensional vector space, and the Morse index $m(i)$ provides a grading. We can define a differential $d : C \to C$ by

$$d(x_i) = \sum_{j \,|\, m(j) - m(i) = 1} \#(i, j)\, x_j,$$

where $\#(i, j)$ denotes the number of gradient flow lines from x_i to x_j, counted with the appropriate signs. This means the following. Choose a Riemannian metric on X, so that the gradient ∇f is a vector field. The downward gradient flow from x_i and the upward gradient flow from x_j are submanifolds of dimension $m(i)$ and $n - m(j)$ respectively. For a sufficiently general metric, they intersect transversely. The index difference being 1 then implies that they intersect in a finite number of flow lines. Choose an orientation of each downward flow; this induces an orientation of each upward flow. Each flow line from x_i to x_j then acquires a sign by comparing four orientations: those of X, the upward and downward flows, and the flow line itself.

It is easy to check that the choice of orientations makes no significant difference. A much harder fact is that $d^2 = 0$. One has to look at flows between critical points of index difference 2: instead of being parametrized by a finite set (= compact 0-manifold) as above, these are parametrized by a disjoint union of closed intervals (= compact 1-manifold), and the crucial point is that there are 0 points in the boundary, when they are counted with the appropriate signs.

So now we have a chain complex, and can take cohomology in the usual way. The amazing fact is that what we get is naturally isomorphic to the rational cohomology of the manifold X!

Notice that this immediately implies the Arnold conjecture in the time-independent case. For the nondegeneracy is then equivalent to the time-independent Hamiltonian $H : M \rightarrow M$ being a Morse function, and the fixed points of ϕ_1 are the critical points of H.

1.6 Bott–Morse Theory

Morse functions always exist; in fact, they are dense among all smooth functions. Nevertheless, suppose fate has endowed us with some smooth $f : X \rightarrow \mathbf{R}$ which is not a Morse function. Can we still use it to determine the cohomology of X? We could try perturbing f to get a Morse function. But often there is no choice of a perturbation which is practical for calculation.

There is one case where we still get some useful information. This is when f is a *Bott–Morse* function: that is, the critical points are a disjoint union of submanifolds, on whose normal bundles the Hessian is nondegenerate. In other words, near every critical point there exist local coordinates in which f can be expressed as before, except that some of the coordinates may be entirely absent. A good example is the pullback of a Morse function by the projection in a fiber bundle.

In the Bott–Morse case, there exists a spectral sequence whose E^2 term is the cohomology of the critical set, bigraded by the Morse index and the degree of the cohomology. It abuts to the cohomology of X. An easy exercise is to show that, in the original Morse case, this boils down to the cochain complex described before. A harder exercise is to show that, in the example of the previous paragraph, it boils down to the Leray spectral sequence.

1.7 Morse Theory on the Loop Space

Now let's return to our Floer set-up: a symplectomorphism $\phi : M \rightarrow M$. We might as well assume that M is connected. Let the *loop space* LM be the set of all smooth maps from the circle S^1 to M. In the case $\phi = \mathrm{id}$, we will define Floer cohomology to be essentially the Morse cohomology of LM, with a "symplectic action function" F playing the role of the Morse function. The loop space is in some sense a manifold, but it is infinite-dimensional, and the upward and downward flows from the critical sets will both be infinite-dimensional as well, so it is lucky that we are optimists.

What is this function F? Suppose first that $\pi_1(M) = 1$, so that LM is connected too. For any $\ell \in LM$, $\ell : S^1 \rightarrow M$, choose a map $\bar{\ell} : D^2 \rightarrow M$ extending ℓ, where D^2 is the disc, and let $F(\ell) = \int_{D^2} \bar{\ell}^* \omega$. This is only defined modulo the integrals of ω on spheres in M, but we can pass to the covering space $\tilde{L}M$ determined by the quotient $\pi_1(LM) \to \pi_2(M)$, and there F is defined without ambiguity. Indeed, $\tilde{L}M$ can be regarded as the space of loops plus homotopy classes of extensions $\bar{\ell}$.

If $\pi_1(M) \neq 1$, then LM has several components, and if we fix a loop in each, we can extend ℓ to a cylinder agreeing with the fixed loop on the other end, and proceed as before.

As a matter of fact, for general ϕ we can do something similar: let the *twisted loop space* be

$$L_\phi M = \{\ell : \mathbf{R} \to M \mid \ell(t+1) = \phi(\ell(t))\},$$

and fix a twisted loop in each connected component. A path from any twisted loop ℓ to the fixed one is a smooth map $\bar{\ell} : \mathbf{R} \times [0,1] \to M$ satisfying the obvious periodicity and boundary properties, and we define $F(\ell) = \int_{[0,1] \times [0,1]} \bar{\ell}^* \omega$ as one would expect.

This function F is a very natural one. Indeed, we can define a symplectic form Ω on $L_\phi M$ as follows. The tangent space to $L_\phi M$ at ℓ consists of sections of $\ell^* TM$ which are periodic in a suitable sense. Define $\Omega(u,v) = \int_0^1 \omega(u,v)\, dt$. Then the Hamiltonian flow of F is exactly reparametrization of twisted loops by time translations.

Consequently, the critical points are exactly the constant loops: these must take values in the fixed-point set X^ϕ by the definition of the twisted loop space, so the critical set can be identified with X^ϕ. Paths in the twisted loop space are, of course, maps $\mathbf{R} \times \mathbf{R} \to M$ with the appropriate periodicity in the first factor. The gradient flow lines turn out to be exactly the *pseudo-holomorphic* maps, that is, maps whose derivatives are linear over $\mathbf{C}$. (For brevity we refer to them henceforth as holomorphic.) Here the choice of an almost complex structure on M compatible with ω has induced a metric g on M and hence a metric G on $L_\phi M$.

Here is a sketch of why the gradient flows are exactly the holomorphic maps. Let $t + iu$ be coordinates on $\mathbf{R}^2 = \mathbf{C}$. A map $\ell : \mathbf{R}^2 \to M$ is a gradient flow if $\partial\ell/\partial u \in T_\ell L_\phi M$ is dual under the metric G to dF, that is, if for all $\nu \in T_\ell L_\phi M$,

$$G\left(\frac{\partial\ell}{\partial u}, \nu\right) = dF(\nu)$$

or

$$\int_0^1 g\left(\frac{\partial\ell}{\partial u}, \nu\right) dt = \int_0^1 \omega\left(\frac{\partial\ell}{\partial t}, \nu\right) dt.$$

Since $\omega(\mu, \nu) = g(i\mu, \nu)$, this is equivalent to

$$i\frac{\partial\ell}{\partial t} = \frac{\partial\ell}{\partial u},$$

which is the complex linearity of the derivative.

If everything is sufficiently generic, F is a Morse function. Then we can go ahead and define our Morse complex, where the differential d counts holomorphic maps. The key claims are that we can make things sufficiently generic by composing with some exact Hamiltonian flow, that $d^2 = 0$ as in the finite-dimensional case, and that the cohomology we get does not depend on the flow.

In many cases, F is not sufficiently generic, but it is still a Bott–Morse function: that is, the critical points are a union of submanifolds, and the Hessian is (in some infinite-dimensional sense!) nondegenerate on each normal bundle. Then we're going to get our spectral sequence. We presume that the Floer cohomology can be calculated from it: a highly nontrivial presumption, of course! This is not Floer's actual approach, but it is still a good way to think about it.

For example, if $\phi = \mathrm{id}$ again, then there is just one critical submanifold, identified with M itself. Hence the E^2 term of the spectral sequence is supported in a single row, so we immediately conclude that $HF^*(\mathrm{id}) = H^*(M)$, provided that our presumption is correct.

That sounds very nice, but only because we cheated. We neglected to pass to the cover $\tilde{L}M$. Up there, there are many critical submanifolds, all diffeomorphic to M but interchanged by deck transformations $\pi_2(M)$. If the Morse indices are different, the spectral sequence won't be supported in a single row, so we need another argument to ensure that the differentials vanish. This is indeed true, but won't be justified, even heuristically, until we discuss the finite-order case a little later on.

So a more truthful statement is that $HF^*(\mathrm{id})$ is a direct sum of many copies of $H^*(M, \mathbf{C})$, one for each element of $\pi_2(M)$. This is conveniently written by introducing $\Lambda = \mathbf{C}[\pi_2(M)]$, the group algebra of $\pi_2(M)$. For example, if $\pi_2(M) \cong \mathbf{Z}$, then $\Lambda \cong \mathbf{C}[q, q^{-1}]$. More generally, there will be variables $q_1, q_2, \ldots$ corresponding to generators $\beta_1, \beta_2, \ldots$ of $\pi_2(M)$. Then we have an isomorphism $HF^*(\mathrm{id}) \cong H^*(M, \Lambda)$. We've glossed over the correct definition of the index, but suffice it to say that the correct grading of $q_i \in H^0(M, \Lambda)$ is $c_1(TM)[\beta_i]$. Here the almost complex structure on M has made the tangent bundle TM into a complex vector bundle.

1.8 Re-Interpretation #1: Sections of the Symplectic Mapping Torus

If you don't like the periodicity condition on our holomorphic maps, here is another way to look at the flow lines. Let the integers act on $\mathbf{C} \times M$, on the first factor by translation by $\mathbf{Z} \subset \mathbf{C}$, on the second by iterating ϕ. This acts freely and symplectically, so the quotient M_ϕ is a symplectic manifold. It is a bundle over the cylinder whose fiber is M, and it admits a canonical flat connection whose monodromy is ϕ. For that reason we call it the *symplectic mapping torus*.

Fixed points of ϕ precisely correspond to flat sections of this bundle. Gradient flow lines of F correspond to holomorphic sections: indeed, both correspond to periodic maps $\ell : \mathbf{R}^2 \to M$ as in the previous section. And the convergence of a flow line to two given fixed points at its ends corresponds to the convergence of the holomorphic section to two given flat sections as we move toward the two ends of the cylinder.

The *periodic Floer homology* of Hutchings is a generalization of this in the case where M is a surface: one looks not only at fixed points, but at unordered k-tuples fixed by ϕ, and the differential consists of k-valued sections, possibly ramified. It is conjectured to be related to Seiberg–Witten Floer cohomology.

1.9 Re-Interpretation #2: Two Lagrangian Submanifolds

Another flavor of Floer cohomology takes as its data a compact symplectic manifold N and two Lagrangian submanifolds $L_1, L_2 \subset N$. Act on one of them by an exact Hamiltonian flow until L_1 intersects L_2 transversely (exercise: this is possible). Then consider the Morse cohomology of the space of paths from L_1 to L_2.

That is, define chains to be formal linear combinations of points $x_i \in L_1 \cap L_2$. And define a differential as before, but where $\#(i, j)$ now counts holomorphic maps from the strip $[0, 1] \times \mathbf{R}$ to N such that the two ends of the strip converge to x_i and x_j. Once again, the grading is contrived in such a way that, if $m(i) - m(j) = 1$, we expect a finite number of such maps (modulo translations of the strip).

This flavor has to do with Floer's work on 3-manifold topology. For example, given a Heegaard decomposition of a 3-manifold, let N be the space of irreducible flat $SU(2)$-connections on the bounding surface, and let L_1, L_2 be the connections that extend as flat connections over the two handlebodies. This satisfies the conditions of the previous paragraph except that N is not compact. Optimistically ignoring this technicality, we may state the *Atiyah–Floer conjecture* which claims that the symplectic Floer cohomology of this N agrees with the *instanton Floer cohomology* of the 3-manifold, also defined by Floer. We won't discuss it here except to say that it is roughly the Morse cohomology of the Chern–Simons function on the space of connections on the 3-manifold.

But we digress. Let's see how the previous flavor of Floer cohomology can be regarded as a special case of this one. Just take $N = M \times M$ with the symplectic form $\pi_1^*\omega - \pi_2^*\omega$ where π_1, π_2 are projections, and let L_1, L_2 be the diagonal and the graph of ϕ. The minus sign is chosen so that these will be Lagrangian. To see how the holomorphic curves in the two alternatives correspond, start with a section of the symplectic mapping torus, project the cylinder 2:1 onto a strip $[0, 1] \times \mathbf{R}$ (branched over the boundary components $0 \times \mathbf{R}$ and $1 \times \mathbf{R}$), trivialize the mapping torus in the natural way over the complement of $1 \times \mathbf{R}$, and define a map $[0, 1] \times \mathbf{R} \to M \times M$ taking a point on the strip to the values of the section at the two points of the cylinder above it, relative to this trivialization. An explicit formula is easy to write down, but why bother?

1.10 Product Structures

Both of the alternatives above suggest a way to introduce a product structure on Floer cohomology. In fact, what we're going to define is a linear functional on

$$HF^*(\phi_1) \otimes HF^*(\phi_2) \otimes HF^*(\phi_3)$$

for any symplectomorphisms satisfying $\phi_1\phi_2\phi_3 = \mathrm{id}$. (Technical detail: since infinitely many powers of q may appear in this element, we may have to pass to a slightly larger coefficient ring $\bar{\Lambda}$, the *Novikov ring*. For example, if $\Lambda = \mathbf{C}[q, q^{-1}]$, then $\bar{\Lambda} = \mathbf{C}[[q]][q^{-1}]$.)

Notice that the chains defining Floer cohomology for ϕ and ϕ^{-1} are formal linear combinations of the same fixed points. If one uses the Kronecker delta to define a nondegenerate pairing between these chains, this descends to a nondegenerate pairing $HF^*(\phi) \otimes HF^*(\phi^{-1}) \to \mathbf{C}$. The linear functional above can then be regarded as a linear map

$$HF^*(\phi_1) \otimes HF^*(\phi_2) \longrightarrow HF^*(\phi_1\phi_2).$$

This ought to satisfy some kind of relation like associativity. In particular, for $\phi_1 = \phi_2 = \mathrm{id}$, it ought to define an associative product on $HF^*(\mathrm{id}) = H^*(M)$. For $\phi_1 = \mathrm{id}$, it makes any $HF^*(\phi)$ into a module over $HF^*(\mathrm{id})$. And so on.

In fact, it has been proved that the Floer product on $HF^*(\mathrm{id})$ coincides with the quantum product coming from Gromov–Witten theory. So we can regard each $HF^*(\phi)$ as a module over the quantum cohomology.

Now, what is the linear functional we promised to define? In analogy with alternative #1, it's given by counting sections of a bundle over a sphere minus three points. (The cylinder was a sphere minus two points.) Call this surface S; then $\pi_1(S)$ is free on two generators. Let $M_{\phi_1, \phi_2} = (\tilde{S} \times M)/\pi_1(S)$, where $\tilde{S}$ is the universal cover and $\pi_1(S)$ acts on M via ϕ_1 and ϕ_2. This is a symplectic bundle over S with fiber M. Now count holomorphic curves asymptotic to fixed points x_j, x_k, x_ℓ of ϕ_1, ϕ_2, ϕ_3 on the three ends.

One has to prove that this induces a homomorphism of complexes. The proof is supposed to be a gluing argument. So is the proof of associativity. The idea is to take a sphere minus three (resp. four) discs, and shrink a loop encircling one (resp. two) of the discs to a point. Then study the limiting behavior of holomorphic sections of the bundles with this base and fiber M as the loop shrinks.

By the way, how can all this be phrased in terms of alternative #2? It's easy to convince yourself that the product functional counts holomorphic triangles in $M \times M$ whose edges lie on the graphs of id, ϕ_1, and $\phi_1\phi_2$. More generally, if

$HF^*(L_1, L_2)$ denotes the Floer cohomology of two Lagrangian submanifolds, then there is supposed to be a product operation

$$HF^*(L_1, L_2) \otimes HF^*(L_2, L_3) \longrightarrow HF^*(L_1, L_3)$$

which counts holomorphic triangles with edges in L_1, L_2, L_3.

From either point of view, it's clear that there is no reason to stop with three punctures. One can include any number, working with a sphere minus n points in alternative 1, or an n-gon in alternative 2, and they will induce $(n-1)$-ary operations on the chain complexes which will descend to Massey products on the cohomology. The compatibility between these operations seems to be what Fukaya is describing in his definition of an A^∞ *category*. There's a substantial literature about complexes equipped with such operations, which it would be quite interesting to apply to Floer cohomology. E.g. the Massey products on a compact Kähler manifold are known to vanish. Is this true of the Floer Massey products?

If you want to go even further, there's no need to insist that S be a punctured sphere: it could be a surface of any genus. Correspondingly, instead of n-gons, you could look at non-simply-connected domains.

1.11 The Finite-Order Case

If ϕ has finite order, say $\phi^k = \mathrm{id}$, then $L_\phi M$ can be regarded as a subspace of LM just by speeding up the path by a factor of k. The symplectic action function on $\tilde{L}M$ restricts to the one on $\tilde{L}_\phi M$, up to a scalar multiple. The Hamiltonian flow of F is reparametrization by time translations, but translations by integer values now act trivially, so the flow induces a circle action. In this situation – when the Hamiltonian flow of F induces a circle action – we say that F is the *moment map* for the action.

Now in finite dimensions, it is well known that moment maps for circle actions are *perfect* Bott–Morse functions, meaning that the differentials in the associated spectral sequence are all zero, or equivalently, that the associated Morse inequalities are equalities. Let's suppose that this remains true in our infinite-dimensional setting. If so, we conclude that *if ϕ has finite order, then*

$$HF^*(\phi) \cong H^*(M^\phi).$$

The author has been informed by Hutchings that, under some technical hypotheses, this result can be proved rigorously. It is, of course, a generalization of Floer's result that $HF^*(\mathrm{id}) = H^*(M)$.

1.12 Givental's Philosophy

Givental's philosophy is that Floer cohomology leads in a natural way to differential equations, and to solutions of those equations. These solutions are in some sense generating functions for numbers of rational curves on M; for example, when M is the quintic threefold, we get the famous Picard–Fuchs equation predicted by mirror symmetry.

Givental considers *equivariant* Floer cohomology (even though this is hard to define rigorously): the circle S^1 acts on LM by rotating the loop. He denotes the generator of $H^*(BS^1)$ by $\hbar$. Every symplectic form ω on M induces an equivariantly closed 2-form p on $\tilde{L}M$. Indeed, with respect to the symplectic form Ω on LM defined earlier, the circle action given by reparametrization is Hamiltonian when we pass to the cover $\tilde{L}M$, and the moment map is exactly the action function F. It is part of the usual package of ideas in equivariant cohomology that $p = \Omega + F$ can be regarded as an equivariantly closed 2-form, the *Duistermaat–Heckman form*.

Suppose for simplicity that M is simply connected. Then $\pi_2(M) = H_2(M, \mathbf{Z})$ by the Hurewicz theorem. If this has rank k, let $\omega_1, \ldots, \omega_k$ be a basis consisting of integral symplectic forms, and let $q_1, \ldots, q_k$ be the deck transformations of $\tilde{L}M$ corresponding to the dual basis of $H_2(M, \mathbf{Z})$. We can act on the Floer cohomology $HF^*(\mathrm{id})$ by multiplication by p_i, or by pullback by q_i. These operations all turn out to commute, except that

$$p_i q_i - q_i p_i = \hbar q_i.$$

The noncommutative algebra D over $\mathbf{C}$ generated by p_i and q_i (and q_i^{-1}, since this is the inverse deck transformation), with these relations, is a familiar one. At any rate, it can be regarded as an algebra of differential operators if we set $q_i = e^{t_i}$ and $p_i = \hbar\, \partial/\partial t_i$.

So we should think of a D-module, such as $HF^*(\mathrm{id})$, as a sheaf on a torus $(\mathbf{C}^\times)^k$ equipped with a connection (or rather, a 1-parameter family of connections parametrized by $\hbar$). As a $\mathbf{C}[q_i]$-module, $HF^*(\mathrm{id})$ is free, so the sheaf is a trivial bundle. Only the connection is nontrivial. What we want to know is encoded in the flat sections of the bundle, which are functions of the q_i (and $\hbar$) with values in $H^*(M)$.

Suppose we are in the good case where $H^*(M)$ is generated by H^2. Then $HF^*(\mathrm{id})$ is a principal D-module generated by $1 \in H^0(M)$: this is plausible, since p_i tends to the cup product with ω_i as $q_i \to 0$. So there is a canonical surjection of D-modules $D \to HF^*(\mathrm{id})$. Its kernel K is generated by a finite number of differential operators, and setting these to zero gives the differential equations that determine what we want to know.

Indeed, knowing the flat sections is the same as knowing $\mathrm{Hom}_D(HF^*(\mathrm{id}), \mathcal{O})$, where $\mathcal{O}$ is the sheaf of regular functions on the torus, for such homomorphisms are just the constant maps in terms of a basis of flat sections. On the other hand, such a thing is also the same as a module homomorphism $D \to \mathcal{O}$ which kills K. It is determined by its value at 1, and this consists of a function which satisfies all the differential equations in K.

This heuristic argument inspired Givental's approach to determining the Gromov–Witten invariants for the quintic threefold, and more generally for Calabi–Yau complete intersections in toric varieties. Instead of using the loop space, he uses spaces of stable maps, which he regards as finite-dimensional approximations to the loop space.

2 Gerbes

And now for something completely different: the definition of a gerbe. The motivation for introducing them is quite simple. We want to consider Floer cohomology with local coefficients in a flat U(1)-bundle over the loop space LM (and its twisted variants). This bundle should of course come from some kind of geometric structure on M, and a U(1)-gerbe will be the best candidate.

Here is the first clue to what a gerbe should be. Isomorphism classes of flat line bundles on LM correspond to $H^1(LM, U(1))$. There is a natural *transgression map* $H^2(M, U(1)) \to H^1(LM, U(1))$ given by taking the Künneth component in $H^1 \otimes H^1$ of the pullback by the evaluation map $LM \times S^1 \to M$. So we might expect gerbes to be objects whose isomorphism classes correspond to $H^2(M, U(1))$.

The good news: such objects exist. They were created in the 1960s by Giraud, who was chiefly interested in nonabelian structure groups. Abelian gerbes were discussed in more detail by Brylinski in a book some 25 years later. The bad news: gerbes rely on the theory of stacks, which we now review in the briefest possible terms.

2.1 Definition of Stacks

Let $\mathcal{T}$ be the category of topological spaces (and continuous maps). The category $\mathcal{S}$ of principal G-bundles (and bundle maps) has an obvious covariant functor to $\mathcal{T}$, namely passing to the base space. It enjoys the following properties:

(a) *Inverses*: any bundle map over the identity $X \to X$ is an isomorphism.

(b) *Pullbacks*: given any P over X and any continuous $f : Y \to X$, there exists Q over Y with a bundle map $Q \to P$, namely the pullback $Q = f^*P$. It is unique up to unique isomorphism, and it satisfies the obvious universal property for bundle maps over some $Z \to X$ factoring through Y.

(c) *Gluing of bundles*: given an open cover U_α of X, bundles P_α over U_α, and isomorphisms $f_{\alpha\beta}$ on the double overlaps (with $f_{\alpha\alpha} := \mathrm{id}$) satisfying $f_{\alpha\beta}f_{\beta\gamma}f_{\gamma\alpha} = \mathrm{id}$ on the triple overlaps, there exists a bundle P over X with isomorphisms g_α over U_α to each P_α satisfying $f_{\alpha\beta}g_\beta = g_\alpha$.

(d) *Gluing of bundle isomorphisms*: given two bundles P, P' over X, an open cover U_α, and isomorphisms from P to P' over each U_α agreeing on the double overlaps, there is a unique global isomorphism from P to P' agreeing with all the given ones. (Note this implies that gluing of bundles is unique up to isomorphism.)

A *stack* over $\mathcal{T}$ is simply any category $\mathcal{S}$, equipped with a covariant functor to $\mathcal{T}$, that satisfies properties (a), (b), (c), (d). Here, of course, "bundle" should be replaced by "object" and "bundle map" by "morphism." In this setting the properties have new, alarming names: (a) and (b) make $\mathcal{S}$ a *category fibered in groupoids*; (c) says that *descent data are effective* and (d) says that *automorphisms are a sheaf*. Notice, by the way, that (c) and (d) implicitly use (b).

You don't really need the base category to be that of topological spaces, of course. It can be any category where the objects are equipped with a Grothendieck topology, such as schemes with the étale topology, which allows us to make sense of open covers.

2.2 Examples of Stacks

(1) The stack of principal G-bundles described above is called the *classifying stack* and denoted BG.

(2) The stack of *flat* principal G-bundles, that is, G-bundles equipped with an atlas whose transition functions are locally constant, with the obvious notion of flat bundle maps. In a flat bundle, nearby fibers (i.e. those in a contractible neighborhood) may be canonically identified.

(3) For a fixed space X, the category whose objects are continuous maps $Y \to X$ and whose morphisms are commutative triangles ending at X. The covariant functor takes a map to its domain. This is a stack, denoted $[X]$ or simply X. Note that in this case (d) becomes trivial, because isomorphism is just equality.

(4) For a topological group G acting on X, the category whose objects lying over Y are pairs consisting of principal G-bundles $P \to Y$ and G-equivariant maps $P \to X$. We leave it to the reader to figure out what the morphisms are. This is a stack, denoted $[X/G]$. Notice that this simultaneously generalizes (1), which is the case $[\cdot/G]$, and (3), which is the case $[X/\cdot]$, where $\cdot$ denotes a point.

(5) A more exotic example: for a fixed line bundle $L \to X$ and a fixed integer n, the category whose objects are triples consisting of a map $f : Y \to X$, a line bundle $M \to Y$, and an isomorphism $M^{\otimes n} \cong f^*L$. This was studied by Cadman, who called it the *stack of nth roots*.

(6) For any two stacks, there is a *Cartesian product* stack whose objects are pairs of objects lying over the same space. For example, an object of $X \times BG$ is a map $Y \to X$ and a principal G-bundle $P \to Y$.

2.3 Morphisms and 2-Morphisms

In the theory of categories, much mischief is caused by our inability to declare that two given objects are equal. In the category of finite-dimensional complex vector spaces, for example, we can't say that $V^{**} = V$. The only accurate statement is that they are naturally isomorphic. So if D is the functor taking a vector space to its dual, we can't say that $DD = \text{id}$. We can only say that there is a natural transformation of functors $DD \Rightarrow$ id. We encounter similar mischief in the theory of stacks.

A *morphism of stacks* $\mathcal{S}' \to \mathcal{S}$ is a functor between categories compatible with the covariant functors to $\mathcal{T}$. A stack equipped with a morphism to $\mathcal{S}$ is called a stack *over* $\mathcal{S}$.

But, if $F, F' : \mathcal{S}' \to \mathcal{S}$ are both morphisms of stacks, we also have the mind-expanding concept of a *2-morphism of morphisms* $\Theta : F \Rightarrow F'$, which is a natural transformation of the corresponding functors. Likewise, a *2-isomorphism of morphisms* is a natural isomorphism of the corresponding functors. For example, if $BGL(n)$ is the stack of (frame bundles of) rank n complex vector bundles, then taking the dual bundle defines a morphism $D : BGL(n) \to BGL(n)$ of stacks, and there exists a 2-isomorphism $DD \Rightarrow$ id.

(Exercise: show that, for X a space and $\mathcal{S}$ a stack, the category of stack morphisms $X \to \mathcal{S}$ and 2-morphisms is equivalent to the category of objects of $\mathcal{S}$ lying over X and morphisms of $\mathcal{S}$ lying over id $: X \to X$.)

(Another exercise: show that the category of automorphisms of BG is equivalent to the category of G-*bitorsors*, that is, G-bundles over a point equipped with a left G-action commuting with the usual right G-action. Hint: define a G-bundle over a stack and observe that any functor on G-bundles over spaces extends canonically to G-bundles over stacks; then consider the image of the tautological G-bundle over BG.)

As a consequence of the mischief, many of the familiar concepts we have in the category of spaces extend to stacks in a more convoluted fashion than one might expect. The basic point is that, instead of just requiring that two objects be equal, we have to choose an isomorphism. We give four key examples.

(1) *The fibered product.* If $\mathcal{R}$ and $\mathcal{R}'$ are stacks with morphisms F and F' to $\mathcal{S}$, the fibered product $\mathcal{R} \times_{\mathcal{S}} \mathcal{R}'$ consists of triples: an object R of $\mathcal{R}$, an object R' of $\mathcal{R}'$, and a choice of an isomorphism $F(R) \to F'(R')$. (Exercise: express Cadman's stack of nth roots as a fibered product. Another exercise: a 2-automorphism of F induces an automorphism of the fibered product.)

(2) *Commutative diagrams.* A diagram of stack morphisms isn't just commutative: we have to make it so by choosing a 2-isomorphism. With a triangle of stacks, for instance, we write the symbol $\Downarrow\Theta$ inside the triangle:

$$\begin{array}{ccc} & \mathcal{R} & \\ \nearrow & \Downarrow\Theta & \searrow \\ \mathcal{Q} & \longrightarrow & \mathcal{S} \end{array}$$

to indicate that there is a 2-isomorphism Θ between the two stack morphisms $\mathcal{Q} \to \mathcal{S}$. When four such triangles with 2-isomorphisms fit together to form a tetrahedron, there is a natural compatibility condition between the 2-isomorphisms (which we leave to the alert reader to work out). If it is satisfied, the tetrahedron is said to be *commutative*.

(3) *Group actions on stacks.* Let Γ be a finite group. A Γ-action on a stack $\mathcal{S}$ consists not only of morphisms $F_\gamma : \mathcal{S} \to \mathcal{S}$ for each $\gamma \in \Gamma$ (with $F_e :=$ id), but also of 2-isomorphisms $\Theta_{\gamma,\gamma'} : F_\gamma F_{\gamma'} \Rightarrow F_{\gamma\gamma'}$ such that, for

any three $\gamma, \gamma', \gamma'' \in \Gamma$, the four 2-isomorphisms $\Theta_{\gamma,\gamma'}$, $\Theta_{\gamma',\gamma''}$, $\Theta_{\gamma\gamma',\gamma''}$, and $\Theta_{\gamma,\gamma'\gamma''}$ form a commutative tetrahedron in the sense alluded to above.

(Exercise: show that the category of Γ-actions on BG is equivalent to the category of extensions of Γ by G. Hint: use the previous exercise.)

(4) *Gluing of stacks.* Let $X = \bigcup X_\alpha$ be a space with an open cover. Recall that a collection of spaces $\pi_\alpha : S_\alpha \to X_\alpha$ may be glued along the open subsets $S_{\alpha\beta} = \pi_\alpha^{-1}(X_\beta)$ using homeomorphisms $f_{\alpha\beta} : S_{\beta\alpha} \to S_{\alpha\beta}$ (with $f_{\alpha\alpha} := \mathrm{id}$), provided that they satisfy $f_{\alpha\beta}f_{\beta\gamma}f_{\gamma\alpha} = \mathrm{id}$ on the triple overlaps. Here is the analogous statement for stacks. If $\mathcal{S}_\alpha$ are stacks over X_α, then they may be glued along $\mathcal{S}_{\alpha\beta} = \mathcal{S}_\alpha \times_X X_\beta$ using isomorphisms $F_{\alpha\beta} : \mathcal{S}_{\beta\alpha} \to \mathcal{S}_{\alpha\beta}$ with $F_{\alpha\alpha} = \mathrm{id}$, provided that there exist 2-isomorphisms $\Theta_{\alpha\beta\gamma} : F_{\alpha\beta}F_{\beta\gamma}F_{\gamma\alpha} \Rightarrow \mathrm{id}$ which in turn form a commutative tetrahedron over the quadruple overlaps.

In the last two examples, the choices of 2-isomorphisms had to satisfy a further condition, namely the commutativity of a tetrahedron. One might ask: why is this adequate? Why isn't some further choice of 3-isomorphisms necessary, and so on? The answer is that categories aren't the most abstract possible structure. In a category, the collection of morphisms between two fixed objects is assumed to be a set. Consequently, it is meaningful to speak of two given 2-isomorphisms as being equal (in contrast to 1-morphisms), since a 2-morphism $F \Rightarrow G$ consists of an element of the set of morphisms $F(C) \to G(C)$ for each object C.

One can, of course, define a more abstract entity, a 2-category, where even the morphisms between fixed objects merely comprise a category. Continuing recursively, one can even define 3-categories, 4-categories, and so on, with their corresponding 2-stacks, 3-stacks, 4-stacks.... Luckily, we will not have to enter this dizzying hall of mirrors.

2.4 Definition of Gerbes

Let's return to the definition of stacks given a while back, and to the principal example BG. This stack actually satisfies two more properties, clearly analogous to (c) and (d):

(c'): *Local existence of bundles*: given any space Y, there is an open cover U_α of Y such that U_α is the base space of a G-bundle.

(d'): *Local existence of bundle isomorphisms*: given two bundles P and P' over Y, there is an open cover U_α such that $P|_{U_\alpha} \cong P'|_{U_\alpha}$.

Of course, (c') could not be more trivial for BG, since the trivial cover and the trivial bundle will do. However, the relative versions of both properties are interesting.

A stack $\mathcal{S}$ over a space X is said to be a *gerbe* over X if:

(c') for any $f : Y \to X$ there is an open cover U_α of Y so that there exists an object of $\mathcal{S}$ lying over each restriction $f|_{U_\alpha}$; and

(d') for any $f : Y \to X$ and any two objects P, P' of $\mathcal{S}$ lying over f, there exists an open cover U_α of Y so that $P|_{U_\alpha} \cong P'|_{U_\alpha}$.

For example, BG is a gerbe over a point. We wish to exhibit nontrivial examples of gerbes over larger spaces.

2.5 The Gerbe of Liftings

To do this, recall first that for any homomorphism $\rho : G \to H$ of Lie groups, one defines the *extension of structure group* of a principal G-bundle $P \to X$ to be the twisted quotient $P_\rho = (P \times H)/G$, where G acts on H via ρ. It is a principal H-bundle over X. As our main example, let $\rho : \mathrm{GL}(n) \to \mathrm{PGL}(n)$ be the projection; then extension by ρ takes a vector bundle to its projectivization. (Here we have intentionally blurred the distinction between the equivalent categories of vector bundles and of frame bundles.)

Now let X be a topological space and P a principal H-bundle. Consider the category of triples consisting of (a) A map $f : Y \to X$; (b) A principal G-bundle $Q \to Y$; (c) An isomorphism $Q_\rho \to f^*P$. It is easily verified that this is a gerbe B over X; call it the *gerbe of liftings* of P. In the main example, Q is a vector bundle over Y whose projectivization is identified with the pullback of a given projective bundle P.

We recognize these triples, don't we? They resemble the triples defining the fibered product two sections back. Indeed, extension of structure group by ρ defines a natural transformation from G-bundles to H-bundles, hence a morphism $B\rho : BG \to BH$; on the other hand, P defines a morphism $X \to BH$, and our gerbe of liftings is nothing but $X \times_{BH} BG$.

It is easiest to understand this gerbe in the case where ρ is surjective, so that we have a short exact sequence

$$1 \longrightarrow A \overset{\sigma}{\longrightarrow} G \overset{\rho}{\longrightarrow} H \longrightarrow 1$$

with A normal. Consider first the case where P is trivialized. Then Q is a principal G-bundle with Q_ρ trivialized, and this precisely means that its structure group is reduced to A, that is, we get a bundle R with $R_\sigma = Q$. Conversely any such R gives Q with Q_ρ trivialized. Hence in this case the gerbe $B \cong X \times BA$.

On the other hand, if P is nontrivial, the gerbe may not be such a product. For example, if P is a projective bundle which does not lift to a vector bundle, then B has no global objects over the identity $X \to X$, but $X \times BA$ does.

In light of all this, a gerbe of liftings is a locally trivial bundle, in the category of stacks, with base X and fiber BA. At least morally speaking, one would like to say that H acts on BA, and that the gerbe is the associated BA-bundle to P. (Exercise: prove this when H is finite. Hint: use the previous exercise.) However, general group actions in the category of stacks turn out to be very slippery.

If the extension of groups is *central*, that is, $A \subset Z(G)$, then things become much simpler, at least at a conceptual level. To begin with, A must be abelian, and "an abelian group is a group in the category of groups," that is, the group operations of multiplication and inversion are group homomorphisms $A \times A \to A$ and $A \to A$, respectively. Consequently, there are good notions of tensor product, and of dual, for A-bundles: namely, the extension of structure group by these homomorphisms. This in turn implies that there are natural morphisms $BA \times BA \to BA$ and $BA \to BA$ making BA into an abelian *group stack* in some sense. The central extension can be regarded as a principal A-bundle over H determining a morphism $H \to BA$, and this morphism is a homomorphism of group stacks. The gerbe of liftings is therefore a principal BA-bundle in this case. However, all this is not as easy to formulate rigorously as it seems, as the precise definition of a group stack is very confusing: associativity need not hold exactly, but only up to 2-isomorphisms which themselves must satisfy compatibility conditions....

2.6 The Lien of a Gerbe

Roughly speaking, an arbitrary gerbe may be described as in the previous section, except that A, instead of being a fixed group, may be a sheaf of groups on X. This sheaf is called the *lien* or *band* of the gerbe. However, since nonabelian phenomena introduce some subtleties, we will discuss only the case analogous to the central extension of the last paragraph. This completely obscures the nonabelian motivation of the founders of the subject, but it is nevertheless sufficient for our purposes.

So let $\mathcal{F}$ be a sheaf of abelian groups over X. An $\mathcal{F}$-*torsor* is a sheaf of sets over X equipped with an atlas of local isomorphisms to $\mathcal{F}$ whose transition functions are given by multiplication by sections of $\mathcal{F}$. An $\mathcal{F}$-*isomorphism* of two $\mathcal{F}$-torsors is an isomorphism of sheaves locally given by multiplication by sections of $\mathcal{F}$.

Hence an $\mathcal{F}$-torsor is acted on by $\mathcal{F}$ itself, and indeed is locally isomorphic to $\mathcal{F}$ as an $\mathcal{F}$-sheaf, but without a choice of an identity element. For example, if $\mathcal{F}$ is the sheaf of continuous functions with values in an abelian group A, then an $\mathcal{F}$-torsor is a principal A-bundle. Or if $\mathcal{F}$ is the sheaf of locally constant functions with values in A, then an $\mathcal{F}$-torsor is a flat A-bundle.

There is a binary operation on $\mathcal{F}$-torsors taking L and L' to $(L \times_X L')/\mathcal{F}$ (with the antidiagonal action), which we denote $L \otimes L'$. For principal A-bundles, it agrees with the tensor product defined before.

Notice, if L, L' are fixed $\mathcal{F}$-torsors, that the sheaf of $\mathcal{F}$-isomorphisms $\mathrm{Isom}(L, L')$ is itself an $\mathcal{F}$-torsor, and the sheaf of $\mathcal{F}$-automorphisms $\mathrm{Aut} L = \mathrm{Isom}(L, L)$ is canonically isomorphic to $\mathcal{F}$ itself.

The collection of all $\mathcal{F}$-torsors forms a stack, indeed a gerbe, $B\mathcal{F}$ over X. More precisely, an object of $B\mathcal{F}$ consists of a map $g : Y \to X$ and a $g^*\mathcal{F}$-torsor.

An $\mathcal{F}$-*gerbe*, then, is defined analogously to a torsor: it is a gerbe over X equipped with an atlas of local isomorphisms to $B\mathcal{F}$ whose transition functions are given by tensor product by sections of $B\mathcal{F}$, that is, torsors on the double overlaps.

For example, the gerbe of liftings of a central extension of H by A is an A-gerbe.

An $\mathcal{F}$-*morphism* of $\mathcal{F}$-gerbes is defined in the obvious way, as is an $\mathcal{F}$-*2-morphism*. To simplify notation, from now on *morphism* will always refer to an $\mathcal{F}$-*morphism* where $\mathcal{F}$-gerbes are concerned, and likewise for 2-morphisms. Note that this is a nontrivial restriction: for example, passage to the dual defines an automorphism of BA, but not an A-automorphism. (Exercise: in terms of the previous exercises, A-automorphisms correspond to the subcategory of bitorsors isomorphic to the trivial one.)

With this convention, an automorphism of an $\mathcal{F}$-gerbe, more or less by definition, is given by $L\otimes$ (that is, tensor product with L) for a fixed $\mathcal{F}$-torsor L. This induces an equivalence of categories, so the 2-morphisms $L\otimes \Rightarrow L'\otimes$ (of gerbe automorphisms) correspond to morphisms $L \to L'$ (of torsors). In particular, the 2-automorphisms $L\otimes \Rightarrow L\otimes$ correspond naturally to sections of $\mathcal{F}$ itself.

We can recover the lien from the gerbe. Suppose we are given a gerbe all of whose objects have abelian automorphism groups. Then the sheaves of automorphisms of any two objects are canonically isomorphic, so they glue together to give a globally defined sheaf $\mathcal{F}$ of abelian groups. It is easy to show that the gerbe is then an $\mathcal{F}$-gerbe. However, if some automorphism groups are nonabelian, this gives rise to the complications ominously alluded to above.

2.7 Classification of Gerbes

At last we are in the position to state a classification result. To avoid complications we confine ourselves to the abelian case, as before.

Theorem (Giraud). The group of isomorphism classes of $\mathcal{F}$-gerbes is isomorphic to $H^2(X, \mathcal{F})$.

Sketch of proof: Trivialize the gerbe on a cover by open sets X_α. The transition functors $F_{\alpha\beta}$ then correspond to $\mathcal{F}$-torsors $L_{\alpha\beta}$ on $X_{\alpha\beta} = X_\alpha \cap X_\beta$. After refining the cover if necessary, we may choose trivializations of these torsors. But, on the triple overlaps $X_{\alpha\beta\gamma}$, we also have the trivializations of $L_{\alpha\beta} \otimes L_{\beta\gamma} \otimes L_{\gamma\alpha}$ given by the 2-isomorphisms $F_{\alpha\beta}F_{\beta\gamma}F_{\gamma\alpha} \Rightarrow \mathrm{id}$. These then determine sections of $\mathcal{F}$ on $X_{\alpha\beta\gamma}$ which constitute a Čech 2-cochain. The tetrahedron condition on the 2-isomorphisms precisely implies that this is closed; and changing the trivializations of the torsors $L_{\alpha\beta}$ adds an exact cocycle.

2.8 Allowing the Base Space to Be a Stack

A general philosophy is that everything that can be done for manifolds should also be attempted for orbifolds. More broadly, everything that can be done for spaces should also be attempted for stacks. In this spirit, we describe here what is meant by a sheaf, a torsor, or a gerbe whose base space is itself a stack. The definition resembles that of a characteristic class.

Let $\mathcal{S}$ be a stack. A *sheaf over* $\mathcal{S}$ is a functor $\mathcal{F}$, over the category of topological spaces, from $\mathcal{S}$ to the category of sheaves. That is, it assigns to every object of $\mathcal{S}$ over Y a sheaf F over Y, and to every morphism of objects over $g : Y \to Y'$ an isomorphism $F \cong g^*F'$. A *torsor* for a given sheaf is defined similarly.

However, we won't define a *gerbe over* $\mathcal{S}$ in the same way, for gerbes (like all stacks) don't just constitute a category, but rather a 2-category. Instead, a gerbe B over $\mathcal{S}$ is a stack over $\mathcal{S}$ such that for all objects of $\mathcal{S}$ over Y, the fibered product $Y \times_\mathcal{S} B$ is a gerbe over Y. An $\mathcal{F}$-gerbe is defined similarly for a sheaf $\mathcal{F}$ over $\mathcal{S}$.

(Exercise: a sheaf of abelian groups over BG corresponds naturally to an abelian group A with a G-action by group automorphisms. A gerbe over BG with lien A corresponds naturally to an extension of G by A so that the action of G on A in the extension is the given one.)

2.9 Definition of Orbifolds

We want to conclude this lecture with a description of the Strominger–Yau–Zaslow proposal for mirror symmetry. To do so, we need two more definitions: of orbifolds and of twisted vector bundles.

First, orbifolds. Roughly speaking, these are stacks locally isomorphic to a quotient of a manifold by a finite group. Readers are cautioned that this definition may differ in a few respects from those in the literature.

Let S be a space. An *orbispace* $\mathcal{S}$ with *coarse moduli space* S is a stack over S so that there exists an open cover $S = \bigcup S_\alpha$ satisfying $S_\alpha \times_S \mathcal{S} \cong [X_\alpha/\Gamma_\alpha]$, where Γ_α is a finite group, and the induced map $X_\alpha/\Gamma_\alpha \to S_\alpha$ of spaces is a homeomorphism. It is an *orbifold* if each X_α is a manifold.

A *smooth structure* on an orbifold is a choice of smooth structure on each X_α so that X_α and X_β induce the same smooth structure on the covering space $X_\alpha \times_\mathcal{S} X_\beta$. (Exercise: this implies that each Γ_α acts smoothly.) A *complex structure* on an orbifold is defined similarly.

2.10 Twisted Vector Bundles

Let B be a gerbe over X with structure group $U(1)$. As we have seen, B is a fiber bundle over X with fiber $BU(1)$. A *twisted vector bundle* for B is a vector bundle over B whose restriction to each fiber is a representation of $U(1)$ (using the last exercise) of pure weight 1.

These are called "twisted" since they can be regarded as locally trivial on open sets $X_\alpha \subset X$, with transition functions $f_{\alpha\beta} : X_{\alpha\beta} \to \mathrm{GL}(n)$. Instead of the usual cocycle condition, we require that $f_{\alpha\beta} f_{\beta\gamma} f_{\gamma\alpha} = b_{\alpha\beta\gamma}$ id where b is a cocycle representative for the isomorphism class of B in $H^2(X, U(1))$.

The same applies to flat gerbes and flat vector bundles.

Twisted vector bundles for a given gerbe clearly form an abelian category, so a *twisted K-theory* may be defined. If the gerbe is trivial, we recover ordinary

K-theory. However, twisted K-theory for a fixed gerbe does not admit a tensor product: rather, we would have to sum over all gerbes, or at least all powers of the fixed one.

2.11 Strominger–Yau–Zaslow

The proposal of Strominger–Yau–Zaslow on mirror symmetry can be described in the language of gerbes and orbifolds. Their remarkable idea is that mirror partners should be Calabi–Yau orbifolds M and $\hat{M}$ of complex dimension n which admit proper maps to the same orbifold Z of real dimension n:

$$
\begin{matrix}
M & & \hat{M} \\
\pi \searrow & & \swarrow \hat{\pi} \\
& Z &
\end{matrix}
$$

so that, if z is a regular value of π and $\hat{\pi}$, the fibers $L_z = \pi^{-1}(z)$ and $\hat{L}_z = \hat{\pi}^{-1}(z)$ are special Lagrangian tori which are in some sense dual to each other. Here *Lagrangian* means Lagrangian with respect to the Kähler form, and *special* means that the imaginary part of the nonzero holomorphic n-form that exists on any Calabi–Yau vanishes on the torus.

The duality between the tori can be required in a strong sense originally envisioned by SYZ, or in a more general sense proposed by Hitchin and involving flat gerbes.

In the original formulation of SYZ, the maps π and $\hat{\pi}$ are assumed to have special Lagrangian sections, giving a basepoint for each L_z and $\hat{L}_z$. This canonically makes them into Lie groups, since a choice of a basis for $T_z^* Z$ determines, via the Kähler form, n commuting vector fields on L_z and $\hat{L}_z$ whose flows define a diffeomorphism to $(S^1)^n$. We then ask for isomorphisms of Lie groups (smoothly depending on z)

$$
\hat{L}_z \cong \mathrm{Hom}(\pi_1(L_z), \mathrm{U}(1))
$$

and vice versa. That is, the tori parametrize isomorphism classes of flat $\mathrm{U}(1)$-bundles on each other.

This formulation was generalized by Hitchin to the case of torus families without sections. It turns out that the absence of a section on M reflects the non-triviality of a gerbe on $\hat{M}$, and vice versa.

So suppose now that M (resp. $\hat{M}$) is equipped with a flat orbifold $\mathrm{U}(1)$-gerbe B (resp. $\hat{B}$) trivial on the fibers of π (resp. $\hat{\pi}$). We can now ask each torus to parametrize isomorphism classes of *twisted* flat $\mathrm{U}(1)$-bundles on the other torus. More than that, we can ask $B|_{L_z}$ to be identified with the stack of twisted flat $\mathrm{U}(1)$-bundles on $\hat{L}_z$, and vice versa. Of course, we want this identification to depend smoothly on $z \in Z$, and we leave it to the reader to specify exactly what this means.

It is extremely difficult to find examples of special Lagrangian tori on Calabi–Yau manifolds. The consensus in the field seems to be that the requirements of SYZ as stated above are too stringent, and that perhaps they must only be satisfied in some limiting sense, say near the "large complex structure limit" in the moduli space of complex structures on the Calabi–Yau. However, the author has studied a few cases where for relatively straightforward reasons (because the metric is, say, hyperkähler or flat) the requirements of SYZ, in the gerbe sense, are seen to be satisfied precisely.

3 Orbifold Cohomology and Its Relatives

What kind of cohomology can be defined for orbifolds? The simplest answer is given in the first section below. Cohomology can be defined for any coefficient ring, or indeed, any sheaf on a stack, in such a way that, if $\mathcal{M}$ is an orbifold with coarse moduli space M,

$$H^*(\mathcal{M}, \mathbf{C}) = H^*(M, \mathbf{C}).$$

However, it has been known for a long time that, for the purposes of string theory, mirror symmetry, and so on, a more refined form of cohomology is preferable. This is the *orbifold cohomology* theory $H^*_{\mathrm{orb}}(\mathcal{M}, \mathbf{C})$, which as a vector space is

$$H^*_{\mathrm{orb}}(\mathcal{M}) = H^*(\mathcal{I}\mathcal{M}).$$

Here $\mathcal{I}\mathcal{M}$ is the so-called *inertia stack*, to be introduced shortly.

We did not specify what coefficient ring to take on the right-hand side, but suppose we choose the Novikov ring from Lecture 1, which is the coefficient ring for Floer cohomology. Then orbifold cohomology admits a quantum cup product whose associativity is a deep and significant fact. Indeed, this is the main reason for studying orbifold cohomology. However, we won't delve into the construction of the product or the proof of associativity. Rather, after defining orbifold cohomology, we will introduce some of its variants and relatives – the version with a flat U(1)-gerbe, for example, and the *Fantechi–Göttsche ring* defined for a global quotient $[X/\Gamma]$ – and then explain how we expect all of these structures to be related to Floer theory.

3.1 Cohomology of Sheaves on Stacks

Just as a sheaf $\mathcal{F}$ on a stack $\mathcal{S}$ is a rule assigning to each object S of $\mathcal{S}$ over Y a sheaf F_S over Y, we can define a *cohomology class* for $\mathcal{F}$ to be a rule assigning to each S an element of $H^*(Y, F_S)$ in a manner compatible with pullbacks. In more fancy categorical language, this is the limit of the functor $H^* \circ \mathcal{F}$ from $\mathcal{S}$ to the category of abelian groups. It is clear that this is a group provided that it is a set! For reasonable sheaves and stacks, this will be true.

For example, if $[X/\Gamma]$ is an orbifold with a sheaf $\mathcal{F}$ regarded as an equivariant sheaf on X, then clearly

$$H^*([X/\Gamma], \mathcal{F}) = H^*(X, \mathcal{F})^\Gamma,$$

where the superscript on the right-hand side denotes the invariant part. If K is a field of characteristic 0, then a theorem of Grothendieck gives a canonical isomorphism

$$H^*(X, K)^{\Gamma} \cong H^*(X/\Gamma, K),$$

so the cohomology of a global quotient (with coefficients in K) coincides with the cohomology of its coarse moduli space.

We can then conclude that the same is true for an arbitrary orbifold $\mathcal{M}$ by using the Mayer–Vietoris spectral sequence. Use a countable atlas where every open set is a global quotient $[X_\alpha/\Gamma_\alpha]$; then the natural map $[X/\Gamma] \to X/\Gamma$ induces isomorphisms $H^*([X_\alpha/\Gamma_\alpha], K) \cong H^*(X_\alpha/\Gamma_\alpha, K)$, and similarly for double overlaps, triple overlaps, and so on. Hence it induces isomorphisms between the double complexes that appear in the Mayer–Vietoris spectral sequences for $\mathcal{M}$ and its coarse moduli space M, and so we conclude that it induces an isomorphism

$$H^*(\mathcal{M}, K) \cong H^*(M, K)$$

when K is a field of characteristic 0.

(Exercise: show that for an arbitrary topological group G and coefficient ring R, there is a natural isomorphism $H^*([X/G], R) \cong H_G^*(X, R)$ where the right-hand side is equivariant cohomology.)

3.2 The Inertia Stack

Let $\mathcal{S}$ be a stack. We can associate to it another stack, the *inertia stack* $\mathcal{IS}$. This is defined to be the stack whose objects over Y are pairs consisting of an object of $\mathcal{S}$ over Y and an automorphism of that object over the identity on Y, and whose morphisms are commutative squares.

If the stack is a quotient by a finite group, the inertia stack can be described explicitly.

Proposition. There is a natural isomorphism

$$\mathcal{I}[X/\Gamma] \cong \bigsqcup_{[\gamma]} [X^\gamma/C(\gamma)],$$

where the disjoint union runs over conjugacy classes in Γ, $X^\gamma = \{x \in X \mid \gamma x = x\}$ is the fixed-point set, and $C(\gamma)$ denotes the centralizer of $\gamma \in \Gamma$.

Sketch of proof. An object of $[X/\Gamma]$ consists of a principal Γ-bundle $P \to Y$ together with a Γ-equivariant map $P \to X$. Hence an object of $\mathcal{I}[X/\Gamma]$ consists of those two things plus an automorphism of P preserving the equivariant map. Since Γ is discrete, any automorphism is given by the right action of some $\gamma \in \Gamma$ commuting with the monodromy group, that is, the image of $\pi_1(Y) \to \Gamma$. Thus the structure group is reduced to $C(\gamma)$, so we get a principal $C(\gamma)$-bundle and an equivariant map to X which, since it is preserved by γ, must have image in X^γ.

It follows directly that, if $\mathcal{M}$ is an orbifold, then so is $\mathcal{J}\mathcal{M}$ (though with components of different dimensions).

(Exercise: prove that there is a natural isomorphism $\mathcal{J}\mathcal{S} \cong \mathcal{S} \times_{\mathcal{S} \times \mathcal{S}} \mathcal{S}$ for any stack $\mathcal{S}$.)

3.3 Orbifold Cohomology

Henceforth, assume that our orbifold $\mathcal{M}$ is *Kähler*, that is, locally $[X_\alpha/\Gamma_\alpha]$ with X_α a Kähler manifold so that X_α and X_β induce the same Kähler structure on the covering space $X_\alpha \times_{\mathcal{M}} X_\beta$. We may then define the *orbifold cohomology* of M to be

$$H^*_{\mathrm{orb}}(\mathcal{M}, \mathbf{C}) = H^*(\mathcal{J}\mathcal{M}, \mathbf{C}).$$

To be more precise, the grading on the orbifold cohomology is not the usual one. Rather, the different connected components have the degrees of their cohomology shifted by different amounts. For a connected component of $[X^\gamma/C(\gamma)] \subset \mathcal{J}[X/\Gamma]$, the so-called *fermionic shift* is defined as follows. Since γ has finite order, it acts on the tangent space $T_x X$ at a point $x \in X^\gamma$ with weights $e^{2\pi i w_1}, \ldots, e^{2\pi i w_n}$ for some rational numbers $w_1, \ldots, w_n \in [0, 1)$. (This is why we need $\mathcal{M}$ Kähler, or at least complex: so that the w_j will be well defined.) Then let $F(\gamma) = \sum_j w_j$. The notation suggests that $F(\gamma)$ is the same on all connected components of $X^\gamma/C(\gamma)$, which is true in most interesting cases. In any case, the grading of the cohomology of the component of $X^\gamma/C(\gamma)$ containing x should be increased by $2F(\gamma)$. For example, the correct grading for $H^*_{\mathrm{orb}}[X/\Gamma]$ is

$$H^k_{\mathrm{orb}}[X/\Gamma] = \bigoplus_{[\gamma]} H^{k-2F(\gamma)}(X^\gamma, \mathbf{C})^{C(\gamma)}.$$

Warning: the fermionic shift may not be an integer! But it will be in many interesting cases, like that of a global quotient $[X/\Gamma]$ provided that the canonical bundle of X has a nowhere vanishing section preserved by Γ (which we might call a *Calabi–Yau orbifold*).

(Exercise: prove that the orbifold Betti numbers of a compact complex orbifold satisfy Poincaré duality. If this is too hard, do it only for $[X/\Gamma]$.)

As we mentioned before, the main interest of orbifold cohomology is that $H^*_{\mathrm{orb}}(\mathcal{M}, \bar{\Lambda}) = H^*_{\mathrm{orb}}(\mathcal{M}, \mathbf{C}) \otimes_{\mathbf{C}} \bar{\Lambda}$ admits an associative quantum product, where $\bar{\Lambda}$ is the Novikov ring from Lecture 1. Indeed, stacks of stable maps to the orbifold $\mathcal{M}$ have been constructed, as discussed in the notes of Abramovich in this volume, and their evaluation maps naturally take values in $\mathcal{J}\mathcal{M}$. So Gromov–Witten invariants provide structure constants for a quantum cup product on $H^*(\mathcal{J}\mathcal{M})$.

There are, of course, algebra homomorphisms $\mathbf{C} \to \bar{\Lambda} \to \mathbf{C}$ (the latter given by taking the constant term), and it is tempting to use these, together with the quantum product on $H^*_{\mathrm{orb}}(\mathcal{M}, \bar{\Lambda})$, to define a product on $H^*_{\mathrm{orb}}(\mathcal{M}, \mathbf{C})$.

This is the so-called *orbifold product*, which in fact slightly predates the orbifold quantum product. It involves only the contributions of stable maps of degree 0. Nevertheless, it usually differs from the standard cup product, as there usually exist stable maps which have degree 0 (indeed, their images in the coarse moduli space are just points) but whose evaluations at different marked points lie in different components of $\mathfrak{IM}$.

3.4 Twisted Orbifold Cohomology

Suppose, now, that we have a flat U(1)-gerbe B on our orbifold $\mathfrak{M}$. This immediately induces a flat U(1)-torsor on $\mathfrak{IM}$. Indeed, each object of $\mathfrak{IM}$ consists of an object of $\mathfrak{M}$ (say over Y) and an automorphism of that object (over id : $Y \to Y$), hence an automorphism of the U(1)-gerbe $Y \times_{\mathfrak{M}} B$ over Y, hence a U(1)-torsor on Y.

Let LB be the flat complex line bundle over $\mathfrak{IM}$ associated to this torsor. Now define the *twisted orbifold cohomology* to be simply

$$H^*_{\mathrm{orb}}(\mathfrak{M}, B) = H^*(\mathfrak{IM}, LB),$$

where the right-hand side refers to cohomology with local coefficients.

The degree should be again adjusted by the fermionic shift, which is the same as before. For a trivial gerbe, we recover the previous notion of orbifold cohomology.

Let's spell out what this is for a global quotient $\mathfrak{M} = [X/\Gamma]$. The line bundle LB over $\mathfrak{IM}$ can be regarded as a collection, indexed by $\gamma \in \Gamma$, of $C(\gamma)$-equivariant line bundles $L_\gamma B$ over X^γ; that is,

$$L_\gamma B = LB|_{[X^\gamma/C(\gamma)]}.$$

Then

$$H^*_{\mathrm{orb}}(\mathfrak{M}, B) = \bigoplus_{[\gamma]} H^*(X^\gamma, L_\gamma B)^{C(\gamma)}.$$

Again, there should be a notion of quantum product on this twisted orbifold cohomology after we tensor with the Novikov ring. What is needed is to show that the flat line bundles agree under the pullbacks to stable map spaces by the relevant evaluation maps.

3.5 The Case of Discrete Torsion

One particularly attractive case has received the most attention in the literature: that of a global quotient $[X/\Gamma]$ with a flat U(1)-gerbe pulled back Γ-equivariantly from a point, that is, a flat U(1)-gerbe pulled back from $B\Gamma$. These are classified, as we saw, by $H^2(B\Gamma, U(1))$. This group is known in the physics literature as the *discrete torsion*, and in the mathematics literature as the *Schur multiplier*. It may be interpreted (and computed) as the group cohomology of Γ with coefficients in the trivial module U(1). It can also be regarded as classifying central extensions

$$1 \longrightarrow \mathrm{U}(1) \longrightarrow \tilde{\Gamma} \longrightarrow \Gamma \longrightarrow 1.$$

What makes such gerbes attractive is, firstly, that they are relatively plentiful: for example, $H^2(\mathbf{Z}_n \times \mathbf{Z}_n, \mathrm{U}(1)) \cong \mathbf{Z}_n$. But also, the flat line bundles $L_\gamma B$ can be calculated over a point and then pulled back to X^γ. Consequently, the underlying line bundles are trivial; only the action of the centralizer $C(\gamma)$ is nontrivial. In the literature, this is sometimes called the *phase*: a homomorphism $C(\gamma) \to \mathrm{U}(1)$.

One can easily show, if $\langle\,,\rangle : \Gamma \times \Gamma \to \mathrm{U}(1)$ is a 2-cocycle representing an element B of discrete torsion in group cohomology, that the phase is given by

$$\delta \mapsto \frac{\langle \gamma, \delta \rangle}{\langle \delta, \gamma \rangle}.$$

Hence the summand $H^*(X^\gamma, L_\gamma B)^{C(\gamma)}$ that appears in the definition of $H^*_{\mathrm{orb}}([X/\Gamma], B)$ is simply the isotypical summand of $H^*(X^\gamma, \mathbf{C})$, regarded as a representation of $C(\gamma)$, that transforms according to the inverse of the phase above.

3.6 The Fantechi–Göttsche Ring

In fact, for a global quotient $[X/\Gamma]$ there is supposed to be a larger ring, equipped with a Γ-action, so that the orbifold cohomology can be recovered as the invariant part. This is the *Fantechi–Göttsche ring*.

Additively it is quite simple: just take

$$HFG^*(X, \Gamma) = \bigoplus_{\gamma \in \Gamma} H^*(X^\gamma, \bar{\Lambda}).$$

Notice that the sum runs over group elements, not just conjugacy classes.

As a representation of Γ it is also quite simple: for each $\delta \in \Gamma$, there is a natural isomorphism $X^\gamma \to X^{\delta\gamma\delta^{-1}}$, hence a pullback on the cohomology that induces an automorphism of $HFG^*(X, \Gamma)$. These fit together to give a Γ-action that acts on the Γ-grading by conjugation.

The nontrivial part is the quantum multiplication. The claim is that there are spaces, akin to those of stable maps, but somehow rigidified so that Γ acts nontrivially on them, and so that the evaluation map goes to $\bigsqcup_\gamma X^\gamma$ instead of just the inertia stack. One should then use these spaces, as in the usual definition of quantum cohomology, to define maps $H^*(X^\gamma, \bar{\Lambda}) \otimes H^*(X^{\gamma'}, \bar{\Lambda}) \to H^*(X^{\gamma\gamma'}, \bar{\Lambda})$.

These spaces, and their virtual classes, are constructed by Fantechi and Göttsche for stable maps of degree 0. As a result, they obtain a ring with degree 0 terms only, whose invariant part carries the orbifold product. But there is every reason to expect a quantum product in this setting.

3.7 Twisting the Fantechi–Göttsche Ring with Discrete Torsion

As the reader may be suspecting, we would like a version of the Fantechi–Göttsche ring which involves a flat unitary gerbe. Let's first indicate how to do this for an element of discrete torsion.

As before, represent our element of discrete torsion by a 2-cocycle $\langle\,,\,\rangle$: $\Gamma \times \Gamma \to U(1)$. Being closed under the differential means that for all $f, g, h \in \Gamma$,

$$\frac{\langle f, g \rangle \langle fg, h \rangle}{\langle f, gh \rangle \langle g, h \rangle} = 1.$$

Now for any two elements $a_g \in H^*(X^g)$ and $b_h \in H^*(X^h)$, regarded as summands of $HFG^*(X, \Gamma)$, we have the usual quantum Fantechi–Göttsche product $a_g \cdot b_h \in H^*(X^{gh})$. Now define a new product by

$$a_g * b_h = \langle g, h \rangle \, a_g \cdot b_h.$$

This need not be commutative or even super-commutative, but it is associative: in fact closedness precisely guarantees this.

The action of Γ on $HFG^*(X, \Gamma)$ given above is no longer a ring homomorphism for the $*$ product. Instead, we need to twist the action as follows: the action of $h \in \Gamma$ takes $H^*(X^g)$ to $H^*(X^{hgh^{-1}})$ by the same map as before, but multiplied by the rather odd factor

$$\frac{\langle h, g \rangle \langle hg, h^{-1} \rangle}{\langle h, h^{-1} \rangle}.$$

The justification for this is that first of all, it now acts by ring homomorphisms for the $*$ product, and second of all, the part invariant under all $h \in \Gamma$ is now twisted orbifold cohomology in the sense defined above.

3.8 Twisting It with an Arbitrary Flat Unitary Gerbe

Next, let's see how the previous section is a special case of putting in an equivariant flat $U(1)$-gerbe. So let B be such a gerbe on X, equivariant under Γ, or equivalently, a gerbe on $[X/\Gamma]$. As before we get a flat line bundle $L_g B$ over X^g, with a lifting of the $C(g)$-action. Additively, we define

$$HFG^*(X, \Gamma; B) = \bigoplus_{g \in \Gamma} H^*(X^g, L_g B),$$

where the terms on the right are cohomology with local coefficients.

As before, to extend the quantum Fantechi–Göttsche product to this twisted case, one would have to show that the flat line bundles agree under the pullbacks, by the relevant evaluation maps, to the spaces akin to those of stable maps. (Exercise: carry this out for degree 0 maps. This amounts to showing that when restricted to $X^{g,h} = X^g \cap X^h$, there is a natural isomorphism $L_{gh}B \cong L_gB \otimes L_hB$.)

There is also, of course, a natural isomorphism induced by $h \in \Gamma$,

$$H^*(X^g, L_gB) \longrightarrow H^*(X^{hgh^{-1}}, L_{hgh^{-1}}B),$$

and so Γ acts on $HFG^*(X, \Gamma; B)$, and the invariant part is the twisted orbifold cohomology. Let's check that, in the case when B is discrete torsion, this isomorphism is simply the one induced by the identification $X^g \to X^{hgh^{-1}}$, times the rather odd factor.

Let $\pi : \tilde{\Gamma} \to \Gamma$ be the central extension determined by B. The automorphism of the category $B\mathrm{U}(1)$ induced by multiplication by g is, of course, just tensorization by the $\mathrm{U}(1)$-torsor $\pi^{-1}(g)$: that is, there is a canonical isomorphism $L_gB \cong \pi^{-1}(g)$. Now, once a cocycle representative is chosen for B, we can identify $\tilde{\Gamma}$ with the product $\Gamma \times \mathrm{U}(1)$, and hence $\pi^{-1}(g)$ with $\mathrm{U}(1)$, but this does not respect the group operation. Nevertheless, let's write $\{g, t\}$ for an element of $\Gamma \times \mathrm{U}(1) = \tilde{\Gamma}$. In terms of this, the map $\pi^{-1}(g) \to \pi^{-1}(hgh^{-1})$ is given by conjugation by any element of $\pi^{-1}(h)$, so we might as well take $\{h, 1\}$. Then what we need to compute is

$$\begin{aligned}
\{h, 1\}\{g, t\}\{h, 1\}^{-1} &= \{h, 1\}\{g, t\}\{h^{-1}, 1/\langle h, h^{-1}\rangle\} \\
&= \{hg, \langle h, g\rangle t\}\{h^{-1}, 1/\langle h, h^{-1}\rangle\} \\
&= \left\{hgh^{-1}, \frac{\langle h, g\rangle \langle hg, h^{-1}\rangle}{\langle h, h^{-1}\rangle} t\right\}
\end{aligned}$$

which shows that, in terms of the identification of $\tilde{\Gamma}$ as a product, the action of h multiplies $\pi^{-1}(g)$ by the rather odd factor.

3.9 The Loop Space of an Orbifold

It is high time to explain what all of these rings are supposed to have to do with Floer cohomology. The claim is that each one can be realized as the Morse cohomology of a symplectic action function on an appropriate analogue of the loop space. Associated to each flat $\mathrm{U}(1)$-gerbe will be a flat line bundle on the loop space, and we should take Floer cohomology with local coefficients. The multiplications defined on orbifold and Floer cohomology should then coincide.

Let's begin with the untwisted Fantechi–Göttsche cohomology. Here, all the pieces are already in place. Observe that, since Γ is a finite group, every element acts on X as a finite-order symplectomorphism, so according to the conjecture from Lecture 1, additively

$$HFG^*(X, \Gamma) = \bigoplus_{\gamma \in \Gamma} H^*(X^\gamma, \bar{\Lambda}) = \bigoplus_{\gamma \in \Gamma} HF^*(\gamma).$$

But both sides also carry a product: on the right, this is thanks to the linear map

$$HF^*(\gamma) \otimes HF^*(\gamma') \longrightarrow HF^*(\gamma\gamma')$$

discussed in Lecture 1. The conjecture is that *this Floer product agrees with the quantum Fantechi–Göttsche product.*

To express this in terms of loop spaces, let $L_\Gamma X = \bigsqcup_{\gamma \in \Gamma} L_\gamma X$, where on the right-hand side γ is regarded as a symplectomorphism of X and $L_\gamma X$ is the twisted loop space as in Lecture 1. The group Γ acts on $L_\Gamma X$ by $\delta \cdot \ell = \delta\ell$; this takes $L_\gamma X$ to $L_{\delta\gamma\delta^{-1}} X$. But since $\gamma^m = \mathrm{id}$ for $m = |\Gamma|$, there is also an action of $S^1 = \mathbf{R}/m\mathbf{Z}$ by translating the parameter. We refer to this as *rotating* the twisted loops. This action commutes with that of Γ, and its moment map is exactly the symplectic action function. The fixed-point set is $\bigsqcup_{\gamma \in \Gamma} X^\gamma$. So the Fantechi–Göttsche ring is supposed to be the Morse cohomology of $L_\Gamma X$ with respect to the action function, which is a perfect Bott–Morse function.

One relatively tractable aspect of this conjecture should be the grading. We have explained how the Fantechi–Göttsche ring is graded: by the usual grading on cohomology corrected by the fermionic shift. On the other hand, Floer cohomology also carries a grading. Under the proposed isomorphism, these gradings presumably agree. Recall, though, that the fermionic shift can be fractional: this is already the case for the obvious action of $\mathbf{Z}_n$ on the Riemann sphere. We artfully evaded discussing the Floer grading for nontrivial symplectomorphisms, but it evidently would have to take account of this.

Now, let's move on to consider orbifold cohomology. For an orbifold $\mathcal{M}$, the space of maps $S^1 \to \mathcal{M}$, in the sense of stacks, can be regarded as a stack $L\mathcal{M}$ in a natural way. An object of $L\mathcal{M}$ over Y is, of course, nothing but an object of $\mathcal{M}$ over $Y \times S^1$. Indeed, we wish to regard $L\mathcal{M}$ as an infinite-dimensional symplectic orbifold, just as the loop space of a manifold is an infinite-dimensional symplectic manifold. We won't attempt to justify this beyond observing that, for a global quotient, we have $L[X/\Gamma] = (L_\Gamma X)/\Gamma$.

Once again a circle acts on $L\mathcal{M}$, and now the fixed-point stack of the circle action can be identified with the inertia stack $\mathcal{I}\mathcal{M}$. Again, these statements have to be considered imprecise since we haven't defined circle actions on stacks. But it is clear what we mean in the case of a global quotient, where we just have a circle action commuting with the Γ-action, and the inertia stack is

$$\mathcal{I}[X/\Gamma] = \left(\bigsqcup_{\gamma \in \Gamma} X^\gamma \right) \Big/ \Gamma.$$

So our claim is, once again, that we should regard $H^*_{\mathrm{orb}}(\mathcal{M})$ as the Morse cohomology of $L\mathcal{M}$, and that the product structures in the Floer and orbifold settings should coincide. Here new technical obstacles would present themselves, for we are asking to do Morse theory on an orbifold, which is problematic even in finite dimensions.

Nevertheless, at a heuristic level, our claims are certainly very plausible. Both the Floer and orbifold cohomologies are defined by "counting" holomorphic maps from a thrice-punctured sphere to the orbifold. The difference lies in what we do to make this formal definition into a mathematically rigorous count. In algebraic geometry, one has the machinery of Gromov–Witten theory, with virtual classes and so on.

In symplectic geometry, on the other hand, one has to perturb the equations and their solutions. As we saw, with a single symplectomorphism ϕ, to define $HF^*(\phi)$ one should perturb with exact Hamiltonians until the fixed points are isolated. For three symplectomorphisms satisfying $\phi_1\phi_2\phi_3 = \mathrm{id}$, to define the map $HF^*(\phi_1) \otimes HF^*(\phi_2) \to HF^*(\phi_3^{-1})$ one should presumably perturb all three simultaneously so that their product remains trivial, but so that all three have isolated fixed points. Any map from a thrice-punctured sphere to a global quotient has monodromy of this form, so this indicates how to define the Floer product on a global quotient. On a general orbifold the situation is not so clear. However, Gromov–Witten invariants of orbifolds have been defined in the symplectic literature.

3.10 Addition of the Gerbe

Now suppose that M, a compact Kähler orbifold, carries a flat U(1)-gerbe B.

Consider a map $\ell : S^1 \to M$. This induces a flat U(1)-gerbe ℓ^*B on S^1. This in turn induces a flat U(1)-gerbe on the universal cover $\mathbf{R}$, together with an automorphism covering the translation $t \mapsto t + 1$. But any flat U(1)-gerbe on $\mathbf{R}$ is trivial, and the trivialization determines another automorphism covering $t \mapsto t + 1$. Comparing the two gives a U(1)-torsor over a point.

The same construction works in families, so any map $S^1 \times Y \to M$ determines a flat U(1)-torsor over Y. In particular, there is a flat U(1)-torsor LB over LM. The isomorphism class of LB is the image of the isomorphism class of B under the transgression map $H^2(M, \mathrm{U}(1)) \to H^1(LM, \mathrm{U}(1))$ defined at the beginning of Lecture 2.

Now, let T be a *trinion*, a sphere minus three disjoint disks, and consider a map $T \to M$. Again this induces a flat U(1)-gerbe on the universal cover $\tilde{T}$, but now (since $\pi_1(T)$ has three generators whose product is 1) this leads to three automorphisms f_1, f_2, f_3 of the trivial gerbe on a point and a 2-isomorphism $f_1 f_2 f_3 \Rightarrow \mathrm{id}$. The 2-isomorphism induces a trivialization of the tensor product $L_1 \otimes L_2 \otimes L_3$ of the three torsors coming from the boundary components.

Again this works in families, so if Y is any space of maps from the trinion to M, we get a trivialization of $\mathrm{ev}_1^* LB \otimes \mathrm{ev}_2^* LB \otimes \mathrm{ev}_3^* LB$, where the *evaluation maps* $\mathrm{ev}_i : Y \to LM$ are given by restriction to the boundary circles. This is why the quantum product makes sense with local coefficients in LB: when we pull back classes by ev_1 and ev_2 and cup them together, they push forward under ev_3 to a class with the appropriate local coefficients. (Note that reversing the orientation of a circle will dualize the relevant torsor.)

3.11 The Non-Orbifold Case

Let's see how this plays out in the case where M is simply a compact Kähler manifold. The isomorphism classes of gerbes then sit in the long exact sequence

$$H^2(M, \mathbf{Z}) \to H^2(M, \mathbf{R}) \to H^2(M, \mathrm{U}(1)) \to H^3(M, \mathbf{Z}) \to H^3(M, \mathbf{R}).$$

The map from integral to real cohomology has as kernel the torsion classes and as image a full lattice, so this boils down to

$$0 \longrightarrow \frac{H^2(M, \mathbf{R})}{H^2(M, \mathbf{Z})} \longrightarrow H^2(M, \mathrm{U}(1)) \longrightarrow \mathrm{Tors}\, H^3(M, \mathbf{Z}) \longrightarrow 0,$$

which of course splits, though not canonically. Consider first what happens as the gerbe B ranges over the torus $H^2(M, \mathbf{R})/H^2(M, \mathbf{Z})$. In this case the following notation is convenient: for any $\beta \in H_2(M, Z)$, write $B^\beta = \exp 2\pi i B(\beta) \in \mathrm{U}(1)$. The torsor LB restricted to the constant loops $M \subset LM$ is, of course, canonically trivial. But, if $F : T \to M$ is any map from the trinion to M taking the boundary circles to constant loops, the trivialization of $\mathrm{ev}_1^* LB \otimes \mathrm{ev}_2^* LB \otimes \mathrm{ev}_3^* LB$ does not agree with the canonical one. Rather, as is easily checked, they differ by the scalar factor B^β, where $\beta = F_*[T]$ is the homology class of F (well defined since F is constant on boundary components).

This introduces an additional weighting factor of B^β in the contributions of degree β holomorphic maps $T \to M$ to the Floer product. Since these are already weighted by q^β, we conclude that *the Floer products parametrized by $B \in H^2(M, \mathbf{R})/H^2(M, \mathbf{Z})$ can all be obtained from the usual one by the change of variables $q \mapsto Bq$.*

In fact, this story extends to the full group $H^2(M, \mathrm{U}(1))$, including $\mathrm{Tors}\, H^3(M, \mathbf{Z})$. For by the universal coefficient theorem $H^2(M, \mathrm{U}(1)) = \mathrm{Hom}(H_2(M, \mathbf{Z}), \mathrm{U}(1))$, so any element whatsoever of $H^2(M, \mathrm{U}(1))$ can be used to introduce a weighting factor on the homology classes of holomorphic maps. Nontrivial torsion in $H^3(M, \mathbf{Z})$ is equivalent to nontrivial torsion in $H_2(M, \mathbf{Z})$ and can be used to provide additional new weightings.

So in the non-equivariant case gerbes do not produce any real novelty. We just recover the usual family of weighting factors on homology classes of stable maps given us by quantum cohomology. This is not really surprising: the gerbe was supposed to produce local systems on LM, but then we passed to a cover $\tilde{L}M$ which trivialized those local systems. However, in the equivariant case we do get something new, namely the twisted quantum products.

3.12 The Equivariant Case

Much as before, if B is a $\mathrm{U}(1)$-gerbe on X, $\phi : X \to X$ a symplectomorphism, and an isomorphism $\phi^* B \cong B$ is given, then a $\mathrm{U}(1)$-torsor $L_\phi B$ is naturally induced on the twisted loop space $L_\phi X$. Now it is no longer true that the restriction of $L_\phi B$ to the constant loops $X^\phi \subset L_\phi X$ must be trivial.

On the loop space $L[X\Gamma] = (L_\Gamma X)/\Gamma$ of a global quotient, then, we get a torsor LB extending the torsor on the inertia stack discussed before. The same thing is presumably true for an orbifold $\mathcal{M}$ that is not a global quotient. For any space of maps from the trinion to $\mathcal{M}$, there should be a trivialization of $\mathrm{ev}_1^* LB \otimes \mathrm{ev}_2^* LB \otimes \mathrm{ev}_3^* LB$, and this should allow a twisted Floer product to be defined. At this point it should be clear: we conjecture that *this agrees with the twisted orbifold quantum product.*

An intriguing question: for the Lagrangian-intersection flavor $HF^*(L_1, L_2)$ of Floer cohomology, is there any analogous way to put in a gerbe?

3.13 A Concluding Puzzle

A basic theorem in K-theory asserts that, on a compact manifold X, the Chern character induces an isomorphism

$$K(X) \otimes \mathbf{C} \cong H^*(X, \mathbf{C}).$$

If a finite group Γ acts on X, then there is a similar theorem for the equivariant K-theory:

$$K_\Gamma(X) \otimes \mathbf{C} \cong \bigoplus_{[\gamma]} H^*(X^\gamma, \mathbf{C})^{C(\gamma)},$$

where the sum runs over conjugacy classes. The right-hand side is exactly what we have been calling $H^*_{\mathrm{orb}}(X/\Gamma, \mathbf{C})$. This can also be made a ring isomorphism, provided that the product structure is appropriately defined on both sides. But it seems to be complicated: the usual product on K-theory goes over to the usual product on the cohomology of the inertia stack (not the orbifold cohomology), so to get a ring homomorphism to orbifold cohomology we have to adjust the operation on K-theory, which we might prefer not to do.

As we have discussed, both sides can be generalized by twisting with a Γ-equivariant gerbe B, so we might hope for something like

$$K_\Gamma(X, B) \otimes \mathbf{C} \cong \bigoplus_{[\gamma]} H^*(X^\gamma, L_\gamma B)^{C(\gamma)}.$$

But now the natural multiplicative structures on the two sides are of completely different types. The twisted K-theory on the left-hand side is a module over the untwisted K-theory $K_\Gamma(X)$, while the right-hand side is a ring in its own right. Can these two algebraic structures be related in any reasonable way?

4 Notes on the Literature

4.1 Notes to Lecture 1

Although it is a textbook that does not purport to give all technical details, the best source for further reading on Floer homology is: D. McDuff and D.A. Salamon, *J-holomorphic curves and symplectic topology*, AMS, 2004, referenced hereinafter as McDuff–Salamon. This is a greatly expanded version of *J-holomorphic curves and quantum cohomology*, AMS, 1994. The same authors have also written a wider survey of symplectic geometry: *Introduction to symplectic topology*, Oxford, 1998.

The Hamiltonian Formalism: See V.I. Arnold, *Mathematical methods of classical mechanics*, Grad. Texts in Math. 60, Springer, 1989, or V. Guillemin and S. Sternberg, *Symplectic techniques in physics*, Cambridge, 1984.

The Arnold Conjecture: Floer's original papers are Morse theory for Lagrangian intersections, *J. Differential Geom.* 28 (1988) 513–547; The unregularized gradient flow of the symplectic action, *Comm. Pure Appl. Math.* 41 (1988) 775–813; Witten's complex and infinite-dimensional Morse theory, *J. Differential Geom.* 30 (1989) 207–221; Symplectic fixed points and holomorphic spheres, *Comm. Math. Phys.* 120 (1989) 575–611. The monotone hypothesis, a technical condition on the first Chern class of the tangent bundle, was removed by H. Hofer and D.A. Salamon, Floer homology and Novikov rings, *The Floer memorial volume*, Progr. Math. 133, Birkhäuser, 1995, and by G. Liu and G. Tian, Floer homology and Arnold conjecture, *J. Differential Geom.* 49 (1998) 1–74. For the Lefschetz fixed-point formula, see Sect. 11.26 of R. Bott and L.W. Tu, *Differential forms in algebraic topology*, Grad. Texts in Math. 82, Springer, 1982.

Morse Theory: The classic reference is J. Milnor, *Morse theory*, Princeton, 1963. The point of view in which the differential counts flow lines did not become popular until the 1980s; for a winsome account from that era, see R. Bott, Morse theory indomitable, *Publ. Math. IHES* 68 (1988) 99–114.

Bott–Morse Theory: The spectral sequence was introduced by Bott in An application of the Morse theory to the topology of Lie-groups, *Bull. Math. Soc. France* 84 (1956) 251–281. See the author's A perfect Morse function on the moduli space of flat connections, *Topology* 39 (2000) 773–787 for a concise account. A thorough discussion of Bott–Morse theory is in D.M. Austin and P.J. Braam, Morse-Bott theory and equivariant cohomology, *The Floer memorial volume*, Progr. Math. 133, Birkhäuser, 1995.

Morse Theory on the Loop Space: See Floer's original papers. The Morse index in the Floer theory is called the *Conley-Zehnder* or *Maslov index*: see McDuff–Salamon, Sect. 12.1.

Re-Interpretations: An inspiring exposition on the various forms of Floer homology is by M.F. Atiyah, New invariants of three- and four-dimensional manifolds, *The mathematical heritage of Hermann Weyl*

(Durham, NC, 1987), Proc. Sympos. Pure Math. 48, AMS, 1988. Another is by J.-C. Sikorav, Homologie associée à une fonctionnelle (d'après A. Floer), *Astérisque* 201-203 (1991) 115–141. For the periodic Floer homology of Hutchings, see M. Hutchings, An index inequality for embedded pseudoholomorphic curves in symplectizations, *J. Eur. Math. Soc.* 4 (2002) 313–361, or M. Hutchings and M. Sullivan, The periodic Floer homology of a Dehn twist, *Algebr. Geom. Topol.* 5 (2005) 301–354.

Product Structures: Proofs that the Floer product on $HF^*(\mathrm{id})$ coincides with the quantum product are given by S. Piunikhin, D. Salamon, and M. Schwarz, Symplectic Floer-Donaldson theory and quantum cohomology, *Contact and symplectic geometry (Cambridge, 1994)*, Cambridge, 1996, and by G. Liu and G. Tian, On the equivalence of multiplicative structures in Floer homology and quantum homology, *Acta Math. Sin. (Engl. Ser.)* 15 (1999) 53–80.

There are no details in the literature of the product structures for arbitrary symplectomorphisms. But there is a sketch in McDuff–Salamon, Sect. 12.6. And the case where M is a Riemann surface has been the subject of several papers, e.g. R. Gautschi, Floer homology of algebraically finite mapping classes, *J. Sympl. Geom.* 1 (2003) 715–765, and P. Seidel, The symplectic Floer homology of a Dehn twist, *Math. Res. Lett.* 3 (1996) 829–834. For the Novikov ring, see McDuff–Salamon Sect. 11.1. For the Fukaya category, see many of Fukaya's papers such as K. Fukaya, Floer homology and mirror symmetry I, *Winter School on Mirror Symmetry, Vector Bundles and Lagrangian Submanifolds*, AMS/IP Stud. Adv. Math. 23, AMS, 2001, or K. Fukaya and P. Seidel, Floer homology, A_∞-categories and topological field theory, *Geometry and physics (Aarhus, 1995)*, Dekker, 1997.

The vanishing of the Massey products on a Kähler manifold is proved in P. Deligne, P. Griffiths, J. Morgan, and D. Sullivan, Real homotopy theory of Kähler manifolds, *Invent. Math.* 29 (1975) 245–274.

The Finite-Order Case: On moment maps and perfect Bott–Morse functions, see F.C. Kirwan, *Cohomology of quotients in symplectic and algebraic geometry*, Princeton, 1984. On the finite-order case, a clearly relevant paper is that of A.B. Givental, Periodic mappings in symplectic topology, *Funct. Anal. Appl.* 23 (1989) 287–300.

Givental's Philosophy is most fully laid out in Homological geometry and mirror symmetry, *Proceedings of the International Congress of Mathematicians (Zürich, 1994)*, vol. 1, Birkhäuser, 1995. But see also his Equivariant Gromov–Witten invariants, *Internat. Math. Res. Notices* 1996 (1996) 613–663, as well as A.B. Givental and B. Kim, Quantum cohomology of flag manifolds and Toda lattices, *Comm. Math. Phys.* 168 (1995), 609–641.

For the "usual package of ideas in equivariant cohomology," see the elegant exposition of M.F. Atiyah and R. Bott, The moment map and equivariant cohomology, *Topology* 23 (1984) 1–28.

4.2 Notes to Lecture 2

Much of the basic information on stacks is lifted from B. Fantechi, Stacks for everybody, *European Congress of Mathematics (Barcelona, 2000)*, vol. 1, Progr. Math. 201, Birkhäuser, 2001, and from W. Fulton, What is a stack? Lecture notes available from www.msri.org/publications/ln/msri/2002/introstacks/fulton/1/index.html.

Some other readable sources are D. Edidin, B. Hassett, A. Kresch, and A. Vistoli, Brauer groups and quotient stacks, *Amer. J. Math.* 123 (2001) 761-777 and A. Vistoli's appendix to Intersection theory on algebraic stacks and on their moduli spaces, *Invent. Math.* 97 (1989) 613–670. Much more formidable and comprehensive is the book of G. Laumon and L. Moret-Bailly, *Champs algébriques*, Ergebnisse Math. 39, Springer, 2000.

Examples of Stacks: The stack of nth roots is discussed by C. Cadman, Using stacks to impose tangency conditions on curves, *Amer. J. Math.*, to appear.

Morphisms and 2-Morphisms: A good basic reference on the relevant category theory is Appendix A of C. Weibel, *An introduction to homological algebra*, Cambridge, 1994. For bitorsors, the tetrahedron condition, and so on, see the book of L. Breen, On the classification of 2-gerbes and 2-stacks, *Astérisque* 225 (1994). Group actions on stacks are meticulously treated by M. Romagny, Group actions on stacks and applications, *Michigan Math. J.* 53 (2005) 209–236.

Definition of Gerbes and the following four sections: The earliest and most comprehensive treatment of gerbes is in the book of J. Giraud, *Cohomologie non abélienne*, Grund. Math. Wiss. 179, Springer, 1971. Abelian gerbes are readably discussed by J.-L. Brylinski, *Loop spaces, characteristic classes and geometric quantization*, Progr. Math. 107, Birkhäuser, 1993. See also the book of Breen and the paper of Edidin et al. cited above.

Definition of Orbifolds: A good general discussion, delivered with the author's usual quirky charm, appears in Sect. 13 of the samizdat lecture notes of W. Thurston; for some reason this was not included in the version that appeared in book form, but it is available from www.msri.org/communications/books/gt3m. Another approach to orbifolds, more closely related to stacks, is that via groupoids, due to Moerdijk and collaborators; see for example I. Moerdijk, Orbifolds as groupoids: an introduction, *Orbifolds in mathematics and physics (Madison, WI, 2001)* Contemp. Math. 310, AMS, 2002.

Twisted Vector Bundles: See, for example, E. Lupercio and B. Uribe, Gerbes over orbifolds and twisted K-theory, *Comm. Math. Phys.* 245 (2004) 449–489, or A. Adem and Y. Ruan, Twisted orbifold K-theory, *Comm. Math. Phys.* 237 (2003) 533–556.

Strominger–Yau–Zaslow: The original article by A. Strominger, E. Zaslow, and S.T. Yau, Mirror symmetry is T-duality, *Nuclear Phys. B* 479 (1996) 243–259, has spawned a vast literature; we mention only the addition of gerbes (a.k.a. "B-fields") by N.J. Hitchin, Lectures on special Lagrangian submanifolds, *Winter School on Mirror Symmetry, Vector Bundles and Lagrangian Submanifolds,* AMS/IP Stud. Adv. Math. 23, AMS, 2001, and an appealing survey by R. Donagi and T. Pantev, Torus fibrations, gerbes, and duality, preprint. The author's papers giving examples where SYZ is satisfied are M. Thaddeus, Mirror symmetry, Langlands duality, and commuting elements of Lie groups, *Internat. Math. Res. Notices* 2001 (2001) 1169–1193, and T. Hausel and M. Thaddeus, Mirror symmetry, Langlands duality, and the Hitchin system, *Invent. Math.* 153 (2003) 197–229.

4.3 Notes to Lecture 3

A good general reference on quantum cohomology and Gromov–Witten invariants (without orbifolds) is Part 4 of K. Hori et al., *Mirror symmetry,* AMS, 2003. This volume comprises the proceedings of a school run by the Clay Mathematics Institute.

Cohomology of Sheaves on Stacks: A convenient reference for Grothendieck's theorem is I.G. Macdonald, Symmetric products of an algebraic curve, *Topology* 1 (1962) 319–343.

Orbifold Cohomology: The orbifold product (where quantum parameters are set to zero) was introduced by W. Chen and Y. Ruan, A new cohomology theory of orbifold, *Comm. Math. Phys.* 248 (2004) 1–31. But the quantum product, though constructed later, appears to be more fundamental: for this see D. Abramovich, T. Graber, and A. Vistoli, Gromov–Witten theory of Deligne-Mumford stacks, preprint. See also Abramovich's notes in this volume.

Twisted Orbifold Cohomology: Among the many interesting recent works on the subject, we mention only two by Y. Ruan: Discrete torsion and twisted orbifold cohomology, *J. Symplectic Geom.* 2 (2003) 1–24, and Stringy orbifolds, *Orbifolds in mathematics and physics (Madison, WI, 2001),* Contemp. Math. 310, AMS, 2002.

The Case of Discrete Torsion: The seminal physics paper is by C. Vafa and E. Witten, On orbifolds with discrete torsion, *J. Geom. Phys.* 15 (1995), 189–214. In fact a whole book had been written by a mathematician, G. Karpilovsky, *The Schur multiplier,* Oxford, 1987.

The Fantechi-Göttsche Ring was introduced by B. Fantechi and L. Göttsche, Orbifold cohomology for global quotients, *Duke Math. J.* 117 (2003) 197–227. Since they set the quantum parameters to zero, the Γ-invariant part of their ring carries the orbifold product. Their product has not yet been fully extended to a quantum product, but there is some relevant discussion of the necessary rigidification in T. Jarvis, R. Kaufmann, and T. Kimura, Pointed admissible G-covers and G-equivariant cohomological field theories, *Compos. Math.* 141 (2005) 926–978, and in the 2006 Ph.D. thesis of Maciek Mizerski at the University of British Columbia.

The Loop Space of an Orbifold: Gromov–Witten invariants for orbifolds are defined symplectically by W. Chen and Y. Ruan in Orbifold Gromov–Witten theory, *Orbifolds in mathematics and physics (Madison, WI, 2001)*, Contemp. Math. 310, AMS, 2002, and algebraically by D. Abramovich, T. Graber, and A. Vistoli, Gromov–Witten theory of Deligne-Mumford stacks, preprint.

A Concluding Puzzle: For the basic theorem in K-theory, see M.F. Atiyah, *K-theory*, Benjamin, 1967. The equivariant version of the theorem is usually attributed to M.F. Atiyah and G.B. Segal, On equivariant Euler characteristics, *J. Geom. Phys.* 6 (1989) 671–677. However, an alternative lineage for this result is traced by A. Adem and Y. Ruan, Twisted orbifold *K*-theory, *Comm. Math. Phys.* 237 (2003) 533–556. Adem and Ruan also give a ring isomorphism from equivariant K-theory to the cohomology of the inertia stack. The adjusted ring homomorphism going to orbifold cohomology is constructed by T. Jarvis, R. Kaufmann, and T. Kimura, Stringy K-theory and the Chern character, preprint. Another such construction, which extends to twisted K-theory, is given by A. Adem, Y. Ruan, and B. Zhang, A stringy product on twisted orbifold K-theory, preprint.

The Moduli Space of Curves and Gromov–Witten Theory

R. Vakil*

Department of Mathematics, Stanford University
Stanford, CA 94305–2125, USA
vakil@math.stanford.edu

1 Introduction

These notes are intended to explain how Gromov–Witten theory has been useful in understanding the moduli space of complex curves. We will focus on the moduli space of smooth curves and how much of the recent progress in understanding it has come through "enumerative" invariants in Gromov–Witten theory, something which we take for granted these days, but which should really be seen as surprising. There is one sense in which it should not be surprising – in many circumstances, modern arguments can be loosely interpreted as the fact that we can understand curves in general by studying branched covers of the complex projective line, as all curves can be so expressed. We will see this theme throughout the notes, from a Riemann-style parameter count in Sect. 2.2 to the tool of relative virtual localization in Gromov–Witten theory in Sect. 5.

These notes culminate in an approach to Faber's intersection number conjecture using relative Gromov–Witten theory (joint work with Goulden and Jackson [GJV3]). One motivation for this article is to convince the reader that our approach is natural and straightforward.

We first introduce the *moduli space of curves*, both the moduli space of smooth curves, and the Deligne–Mumford compactification, which we will see is something forced upon us by nature, not arbitrarily imposed by man. We will then define certain geometrically natural cohomology classes on the moduli space of smooth curves (the tautological subring of the cohomology ring), and discuss Faber's foundational conjectures on this subring. We will then extend these notions to the moduli space of stable curves, and discuss Faber-type conjectures in this context. A key example is Witten's conjecture,

* Partially supported by NSF CAREER/PECASE Grant DMS–0228011, and an Alfred P. Sloan Research Fellowship.

2000 Mathematics Subject Classification: Primary 14H10, 14H81, 14N35, Secondary 14N10, 53D45, 14H15.

K. Behrend, M. Manetti (eds.), *Enumerative Invariants in Algebraic Geometry and String Theory*. Lecture Notes in Mathematics 1947,
© Springer-Verlag Berlin Heidelberg 2008

which really preceded (and motivated) Faber's conjectures, and opened the floodgates to the last decade's flurry of developments. We will then discuss other relations in the tautological ring (both known and conjectural). We will describe Theorem ⋆ (Theorem 4.1), a blunt tool for proving many statements, and Y.-P. Lee's Invariance Conjecture, which may give all relations in the tautological ring. In order to discuss the proof of Theorem ⋆, we will be finally drawn into Gromov–Witten theory, and we will quickly review the necessary background. In particular, we will need the notion of "relative Gromov–Witten theory", including Jun Li's degeneration formula [Li1, Li2] and the relative virtual localization formula [GrV3]. Finally, we will use these ideas to tackle Faber's intersection number conjecture.

Because the audience has a diverse background, this article is intended to be read at many different levels, with as much rigor as the reader is able to bring to it. Unless the reader has a solid knowledge of the foundations of algebraic geometry, which is most likely not the case, he or she will have to be willing to take a few notions on faith, and to ask a local expert a few questions.

We will cover a lot of ground, but hopefully this article will include enough background that the reader can make explicit computations to see that he or she can actively manipulate the ideas involved. You are strongly encouraged to try these ideas out via the exercises. They are of varying difficulty, and the amount of rigor required for their solution should depend on your background.

Here are some suggestions for further reading. For a gentle and quick introduction to the moduli space of curves and its tautological ring, see [V2]. For a pleasant and very detailed discussion of moduli of curves, see Harris and Morrison's foundational book [HM]. An on-line resource discussing curves and links to topology (including a glossary of important terms) is available at [GiaM]. For more on curves, Gromov–Witten theory, and localization, see [HKKPTVVZ, Chaps. 22–27], which is intended for both physicists and mathematicians. Cox and Katz' wonderful book [CK] gives an excellent mathematical approach to mirror symmetry. There is as of yet no ideal book introducing (Deligne–Mumford) stacks, but Fantechi's [Fan] and Edidin's [E] both give an excellent idea of how to think about them and work with them, and the appendix to Vistoli's paper [Vi] lays out the foundations directly, elegantly, and quickly, although this is necessarily a more serious read.

Acknowledgments. I am grateful to the organizers of the June 2005 conference in Cetraro, Italy on "Enumerative invariants in algebraic geometry and string theory" (Kai Behrend, Barbara Fantechi, and Marco Manetti), and to Fondazione C.I.M.E. (Centro Internazionale Matematico Estivo). I learned this material from my co-authors Graber, Goulden, and Jackson, and from the other experts in the field, including Carel Faber, Rahul Pandharipande, Y.-P. Lee, ..., whose names are mentioned throughout this article. I thank Carel Faber, Soren Galatius, Tom Graber, Arthur Greenspoon, Y.-P. Lee and Rahul Pandharipande for improving the manuscript. I am very grateful to Renzo Cavalieri, Sam Payne, and the participants in their April 2006 "Moduli space of Curves and Gromov–Witten Theory Workshop" at the University of Michigan for their close reading and many detailed suggestions.

2 The Moduli Space of Curves

We begin with some conventions and terminology. We will work over $\mathbb{C}$, although these questions remain interesting over arbitrary fields. We will work algebraically, and hence only briefly mention other important approaches to the subjects, such as the construction of the moduli space of curves as a quotient of Teichmüller space.

By *smooth curve*, we mean a compact (also known as proper or complete), smooth (also known as nonsingular) complex curve, i.e. a Riemann surface, see Fig. 1. Our curves will be connected unless we especially describe them as "possibly disconnected". In general our *dimensions* will be algebraic or complex, which is why we refer to a Riemann surface as a curve – they have algebraic/complex dimension 1. Algebraic geometers tend to draw "half-dimensional" cartoons of curves (see also Fig. 1).

Fig. 1. A complex curve, and its real "cartoon"

The reader likely needs no motivation to be interested in Riemann surfaces. A natural question when you first hear of such objects is: what are the Riemann surfaces? How many of them are there? In other words, this question asks for a classification of curves.

2.1. *Genus.* A first invariant is the *genus* of the smooth curve, which can be interpreted in three ways: (1) the number of holes (*topological genus*; for example, the genus of the curve in Fig. 1 is 3), (2) dimension of space of differentials $(= h^0(C, \Omega_C)$, *geometric genus*), and (3) the first cohomology group of the sheaf of algebraic functions $(h^1(C, \mathcal{O}_C)$, *arithmetic genus*). These three notions are the same. Notions (2) and (3) are related by *Serre duality*

$$(1) \qquad \boxed{H^0(C, \mathcal{F}) \times H^1(C, \mathcal{K} \otimes \mathcal{F}^*) \to H^1(C, \mathcal{K}) \cong \mathbb{C}}$$

where $\mathcal{K}$ is the canonical line bundle, which for smooth curves is the sheaf of differentials Ω_C. Here $\mathcal{F}$ can be any finite rank vector bundle; H^i refers to sheaf cohomology. Serre duality implies that $h^0(C, \mathcal{F}) = h^1(C, \mathcal{K} \otimes \mathcal{F}^*)$, hence (taking $\mathcal{F} = \mathcal{K}$). $h^0(C, \Omega_C) = h^1(C, \mathcal{O}_C)$. (We will use these important facts in the future!)

As we are working purely algebraically, we will not discuss why (1) is the same as (2) and (3).

2.2. There Is a $(3g-3)$-Dimensional Family of Genus g Curves

Remarkably, it was already known to Riemann [R, p. 134] that there is a "$(3g-3)$-dimensional family of genus g curves". You will notice that this can't possibly be right if $g = 0$, and you may know that this isn't right if $g = 1$, as you may have heard that elliptic curves are parametrized by the j-line, which is one-dimensional. So we will take $g > 1$, although there is a way to extend to $g = 0$ and $g = 1$ by making general enough definitions. (Thus there is a "(-3)-dimensional moduli space" of genus 0 curves, if you define moduli space appropriately – in this case as an Artin stack. But that is another story.)

Let us now convince ourselves (informally) that there is a $(3g-3)$-dimensional family of genus g curves. This will give me a chance to introduce some useful facts that we will use later. I will use the same notation for vector bundles and their sheaves of sections. The sheaf of sections of a line bundle is called an *invertible sheaf*.

We will use five ingredients:

(1) *Serre duality* (1). This is a hard fact.

(2) *The Riemann–Roch formula.* If $\mathcal{F}$ is any coherent sheaf (for example, a finite rank vector bundle) then

$$h^0(C, \mathcal{F}) - h^1(C, \mathcal{F}) = \deg \mathcal{F} - g + 1.$$

This is an easy fact, although I will not explain why it is true.

(3) Line bundles of negative degree have no non-zero sections: if $\mathcal{L}$ is a line bundle of negative degree, then $\boxed{h^0(C, \mathcal{L}) = 0}$. Here is why: the degree of a line bundle $\mathcal{L}$ can be defined as follows. Let s be any non-zero meromorphic section of $\mathcal{L}$. Then the degree of $\mathcal{L}$ is the number of zeros of s minus the number of poles of s. Thus if $\mathcal{L}$ has an honest non-zero section (with no poles), then the degree of s is at least 0.

Exercise. If $\mathcal{L}$ is a degree 0 line bundle with a non-zero section s, show that $\mathcal{L}$ is isomorphic to the trivial bundle (the sheaf of functions) $\mathcal{O}$.

(4) Hence if $\mathcal{L}$ is a line bundle with $\deg \mathcal{L} > \deg \mathcal{K}$, then $h^1(C, \mathcal{L}) = 0$ by Serre duality, from which $\boxed{h^0(C, \mathcal{L}) = \deg \mathcal{L} - g + 1}$ by Riemann–Roch.

(5) *The Riemann–Hurwitz formula.* Suppose $C \to \mathbb{P}^1$ is a degree d cover of the complex projective line by a genus g curve C, with ramification $r_1, \ldots, r_n$ at the ramification points on C. Then

$$\chi_{\text{top}}(C) = d\chi_{\text{top}}(\mathbb{P}^1) - \sum (r_i - 1),$$

where χ_{top} is the topological Euler characteristic, i.e.

$$(2) \qquad 2 - 2g = 2d - \sum (r_i - 1).$$

We quickly review the language of divisors and line bundles on smooth curves. A *divisor* is a formal linear combination of points on C, with integer coefficients, finitely many non-zero. A divisor is *effective* if the coefficients are non-negative. The *degree* of a divisor is the sum of its coefficients. Given a divisor $D = \sum n_i p_i$ (where the p_i form a finite set), we obtain a line bundle $\mathcal{O}(D)$ by "twisting the trivial bundle n_i times at the point p_i". This is best understood in terms of the sheaf of sections. Sections of the sheaf $\mathcal{O}(D)$ (over some open set) correspond to meromorphic functions that are holomorphic away from the p_i; and if $n_i > 0$, have a pole of order at most n_i at p_i; and if $n_i < 0$, have a zero of order at least $-n_i$ at p_i. Each divisor yields a line bundle along with a meromorphic section (obtained by taking the function 1 in the previous sentence's description). Conversely, each line bundle with a non-zero meromorphic section yields a divisor, by taking the "divisor of zeros and poles": if s is a non-zero meromorphic section, we take the divisor which is the sum of the zeros of s (with multiplicity) minus the sum of the poles of s (with multiplicity). These two constructions are inverse to each other. In short, line bundles with the additional data of a non-zero *meromorphic* section correspond to divisors. This identification is actually quite subtle the first few times you see it, and it is worth thinking through it carefully if you have not done so before. Similarly, line bundles with the additional data of a non-zero *holomorphic* section correspond to *effective* divisors.

We now begin our dimension count. We do it in three steps.

Step 1. Fix a curve C, and a degree d. Let $\operatorname{Pic}^d C$ be the set of degree d line bundles on C. Pick a point $p \in C$. Then there is an bijection $\operatorname{Pic}^0 C \to \operatorname{Pic}^d C$ given by $\mathcal{F} \to \mathcal{F}(dp)$. (By $\mathcal{F}(dp)$, we mean the "twist of $\mathcal{F}$ at p, d times", which is the same construction sketched two paragraphs previously. In terms of sheaves, if $d > 0$, this means the sheaf of meromorphic sections of $\mathcal{F}$, that are required to be holomorphic away from p, but may have a pole of order at most d at p. If $d < 0$, this means the sheaf of holomorphic sections of $\mathcal{F}$ that are required to have a zero of order at least $-d$ at p.) If we believe $\operatorname{Pic}^d C$ has some nice structure, which is indeed the case, then we would expect that this would be an isomorphism. In fact, Pic^d can be given the structure of a complex manifold or complex variety, and this gives an isomorphism of manifolds or varieties.

Step 2: "$\dim \operatorname{Pic}^d C = g$". There are quotes around this equation because so far, $\operatorname{Pic}^d C$ is simply a set, so this will just be a plausibility argument. By Step 1, it suffices to consider any $d > \deg \mathcal{K}$. Say $\dim \operatorname{Pic}^d C = h$. We ask: how many degree d *effective divisors* are there (i.e. what is the dimension of this family)? The answer is clearly d, and C^d surjects onto this set (and is usually $d!$-to-1).

But we can count effective divisors in a different way. There is an h-dimensional family of line bundles by hypothesis, and each one of these has a $(d - g + 1)$-dimensional family of non-zero sections, each of which gives a divisor of zeros. But two sections yield the same divisor if one is a multiple of the other. Hence we get: $h + (d - g + 1) - 1 = h + d - g$.

Thus $d = h + d - g$, from which $h = g$ as desired.

Note that we get a bit more: if we believe that Pic^d has an algebraic structure, we have a fibration $(C^d)/S_d \to \text{Pic}^d$, where the fibers are isomorphic to $\mathbb{P}^{d-g}$. In particular, Pic^d is reduced (I won't define this!), and irreducible. (In fact, as many of you know, it is isomorphic to the dimension g abelian variety $\text{Pic}^0 C$.)

Step 3. Say $\mathcal{M}_g$ has dimension p. By fact **(4)** above, if $d \gg 0$, and D is a divisor of degree d, then $h^0(C, \mathcal{O}(D)) = d - g + 1$. If we take two general sections s, t of the line bundle $\mathcal{O}(D)$, we get a map to $\mathbb{P}^1$ (given by $p \to [s(p); t(p)]$ – note that this is well-defined), and this map is degree d (the preimage of $[0; 1]$ is precisely $\text{div } s$, which has d points counted with multiplicity). Conversely, any degree d cover $f : C \to \mathbb{P}^1$ arises from two linearly independent sections of a degree d line bundle. (To get the divisor associated to one of them, consider $f^{-1}([0; 1])$, where points are counted with multiplicities; to get the divisor associated to the other, consider $f^{-1}([1; 0])$.) Note that (s, t) gives the same map to $\mathbb{P}^1$ as (s', t') if and only (s, t) is a scalar multiple of (s', t'). Hence the number of maps to $\mathbb{P}^1$ arising from a fixed curve C and a fixed line bundle $\mathcal{L}$ corresponds to the choices of two sections ($2(d - g + 1)$ by fact **(4)**), minus 1 to forget the scalar multiple, for a total of $2d - 2g + 1$. If we let the the line bundle vary, the number of maps from a fixed curve is $2d - 2g + 1 + \dim \text{Pic}^d(C) = 2d - g + 1$. If we let the curve also vary, we see that the number of degree d genus g covers of $\mathbb{P}^1$ is $\boxed{p + 2d - g + 1}$.

But we can also count this number using the Riemann–Hurwitz formula **(2)**. By that formula, there will be a total of $2g + 2d - 2$ branch points (including multiplicity). Given the branch points (again, with multiplicity), there is a finite amount of possible monodromy data around the branch points. The Riemann Existence Theorem tells us that given any such monodromy data, we can uniquely reconstruct the cover, so we have

$$p + 2d - g + 1 = 2g + 2d - 2,$$

from which $\boxed{p = 3g - 3}$.

Thus there is a $(3g - 3)$-dimensional family of genus g curves! (By showing that the space of branched covers is reduced and irreducible, we could again "show" that the moduli space is reduced and irreducible.)

2.3. The Moduli Space of Smooth Curves

It is time to actually define the moduli space of genus g smooth curves, denoted $\mathcal{M}_g$, or at least to come close to it. By "moduli space of curves" we mean a "parameter space for curves". As a first approximation, we mean the set of curves, but we want to endow this set with further structure (ideally that of a manifold, or even of a smooth complex variety). This structure should be given by nature, not arbitrarily defined.

Certainly if there were such a space $\mathcal{M}_g$, we would expect a universal curve over it $\mathcal{C}_g \to \mathcal{M}_g$, so that the fiber above the point $[C]$ representing a curve

C would be that same C. Moreover, whenever we had a family of curves parametrized by some base B, say $\mathcal{C}_B \to B$ (where the fiber above any point $b \in B$ is some smooth genus g curve C_b), there should be a map $f : B \to \mathcal{M}_g$ (at the level of sets sending $b \in B$ to $[C_b] \in \mathcal{M}_g$), and then $f^*\mathcal{C}_g$ should be isomorphic to $\mathcal{C}_B$.

We can turn this into a precise definition. The families we should consider should be "nice" ("fibrations" in the sense of differential geometry). It turns out that the corresponding algebraic notion of "nice" is *flat*, which I will not define here. We can *define* $\mathcal{M}_g$ to be the scheme such that the maps from any scheme B to it are in natural bijection with nice (flat) families of genus g curves over B. (Henceforth all families will be assumed to be "nice" = flat.) Some thought will convince you that only one space (up to isomorphism) exists with this property. This "abstract nonsense" is called *Yoneda's Lemma*. The argument is general, and applies to nice families of any sort of thing. Categorical translation: we are saying that this contravariant functor of families is *represented* by the functor $\mathrm{Hom}(\cdot, \mathcal{M}_g)$. Translation: if such a space exists, then it is unique, up to unique isomorphism.

If there is such a moduli space $\mathcal{M}_g$, we gain some additional information: cohomology classes on $\mathcal{M}_g$ are "characteristic classes" for families of genus g curves. More precisely, given any family of genus g curves $\mathcal{C}_B \to B$, and any cohomology class $\alpha \in H^*(\mathcal{M}_g)$, we have a cohomology class on B: if $f : B \to \mathcal{M}_g$ is the moduli map, take $f^*\alpha$. These characteristic classes behave well with respect to pullback: if $\mathcal{C}_{B'} \to B'$ is a family obtained by pullback from $\mathcal{C}_B \to B$, then the cohomology class on B' induced by α is the pullback of the cohomology class on B induced by α. The converse turns out to be true: any such "universal cohomology class", defined for all families and well-behaved under pullback, arises from a cohomology class on $\mathcal{M}_g$. (The argument is actually quite tautological, and the reader is invited to think it through.) More generally, statements about the geometry of $\mathcal{M}_g$ correspond to "universal statements about all families".

Here is an example of a consequence. A curve is *hyperelliptic* if it admits a 2-to-1 cover of $\mathbb{P}^1$. In the space of smooth genus 3 curves $\mathcal{M}_3$, there is a Cartier divisor of hyperelliptic curves, which means that the locus of hyperelliptic curves is locally cut out by a single equation. Hence in *any* family of genus 3 curves over an arbitrarily horrible base, the hyperelliptic locus are cut out by a single equation. (For scheme-theoretic experts: for any family $\mathcal{C}_B \to B$ of genus 3 curves, there is then a closed subscheme of B corresponding to the hyperelliptic locus. What is an intrinsic scheme-theoretic definition of this locus?)

Hence all we have to do is show that there is such a scheme $\mathcal{M}_g$. Sadly, there is no such scheme! We could just throw up our hands and end these notes here. There are two patches to this problem. One solution is to relax the definition of moduli space (to get the notion of *coarse moduli space*), which doesn't quite parametrize all families of curves. A second option is to extend the notion of *space*. The first choice is the more traditional one, but it is becoming increasingly clear that the second choice the better one.

This leads us to the notion of a *stack*, or in this case, the especially nice stack known as a *Deligne–Mumford stack*. This is an extension of the idea of a scheme. Defining a Deligne–Mumford stack correctly takes some time, and is rather tiring and uninspiring, but dealing with Deligne–Mumford stacks on a day-to-day basis is not so bad – you just pretend it is a scheme. One might compare it to driving a car without knowing how the engine works, but really it is more like driving a car while having only the vaguest idea of what a car is.

Thus I will content myself with giving you a few cautions about where your informal notion of Deligne–Mumford stack should differ with your notion of scheme. (I feel less guilty about this knowing that many analytic readers will be similarly uncomfortable with the notion of a scheme.) The main issue is that when considering cohomology rings (or the algebraic analog, Chow rings), we will take $\mathbb{Q}$-coefficients in order to avoid subtle technical issues. The foundations of intersection theory for Deligne–Mumford stacks were laid by Vistoli in [Vi]. (However, thanks to work of Andrew Kresch [Kr], it is possible to take integral coefficients using the Chow ring. Then we have to accept the fact that cohomology groups can be non-zero even in degree higher than the dimension of the space. This is actually something that for various reasons we *want* to be true, but such a discussion is not appropriate in these notes.)

A smooth (or nonsingular) Deligne–Mumford stack (over $\mathbb{C}$) is essentially the same thing as a complex orbifold. The main caution about saying that they are the same thing is that there are actually three different definitions of orbifold in use, and many users are convinced that their version is the only version in use, causing confusion for readers such as myself.

Hence for the rest of these notes, we will take for granted that there is a moduli space of smooth curves $\mathcal{M}_g$ (and we will make similar assumptions about other moduli spaces).

Here are some *facts* about the moduli space of curves. The space $\mathcal{M}_g$ has (complex) dimension $3g - 3$. It is smooth (as a stack), so it is an orbifold (given the appropriate definition), and we will imagine that it is a manifold. We have informally seen that it is irreducible.

We make a brief excursion outside of algebraic geometry to show that this space has some interesting structure. In the analytic setting, $\mathcal{M}_g$ can be expressed as the quotient of *Teichmüller space* (a subset of $\mathbb{C}^{3g-3}$ homeomorphic to a ball) by a discrete group, known as the *mapping class group*. Hence the cohomology of the quotient $\mathcal{M}_g$ is the group cohomology of the mapping class group. (Here it is essential that we take the quotient as an orbifold/stack.) Here is a fact suggesting that the topology of this space has some elegant structure:

$$(3) \qquad\qquad \chi(\mathcal{M}_g) = B_{2g}/2g(2g - 2)$$

(due to Harer and Zagier [HZ]), where B_{2g} denotes the $2g$th Bernoulli number.

Other exciting recent work showing the attractive structure of the cohomology ring is Madsen and Weiss' proof of Madsen's generalization of

Mumford's conjecture [MW]. We briefly give the statement. There is a natural isomorphism between $H^*(\mathcal{M}_g; \mathbb{Q})$ and $H^*(\mathcal{M}_{g+1}; \mathbb{Q})$ for $* < (g-1)/2$ (due to Harer and Ivanov). Hence we can define the ring we could informally denote by $H^*(\mathcal{M}_\infty; \mathbb{Q})$. Mumford conjectured that this is a free polynomial ring generated by certain cohomology classes (κ-classes, to be defined in Sect. 3.1). Madsen and Weiss proved this, and a good deal more. (See [T] for an overview of the topological approach to the Mumford conjecture, and [MT] for a more technical discussion.)

2.4. Pointed Nodal Curves, and the Moduli Space of Stable Pointed Curves

As our moduli space $\mathcal{M}_g$ is a smooth orbifold of dimension $3g - 3$, it is wonderful in all ways but one: it is not compact. It would be useful to have a good compactification, one that is still smooth, and also has good geometric meaning. This leads us to extend our notion of smooth curves slightly.

A *node* of a curve is a singularity analytically isomorphic to $xy = 0$ in $\mathbb{C}^2$. A *nodal curve* is a curve (compact, connected) smooth away from a finite number of points (possibly zero), which are nodes. An example is sketched in Fig. 2, in both "real" and "cartoon" form. One caution with the "real" picture: the two branches at the node are not tangent; this optical illusion arises from the need of our limited brains to represent the picture in three-dimensional space. A *pointed nodal curve* is a nodal curve with the additional data of n distinct smooth points labeled 1 through n (or n distinct labels of your choice, such as p_1 through p_n).

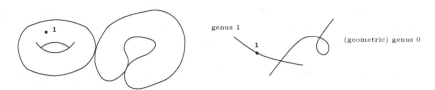

Fig. 2. A pointed nodal curve, and its real "cartoon"

The *geometric genus* of an *irreducible* curve is its genus once all of the nodes are "unglued". For example, the components of the curve in Fig. 2 have genus 1 and 0.

We define the *(arithmetic) genus* of a pointed nodal curve informally as the genus of a "smoothing" of the curve, which is indicated in Fig. 3. More formally, we define it as $h^1(C, \mathcal{O}_C)$. This notion behaves well with respect to deformations. (More formally, it is locally constant in flat families.)

Exercise (for those with enough background): If C has δ nodes, and its irreducible components have geometric genus $g_1, \ldots, g_k$ respectively, show that the arithmetic genus of C is $\sum_{i=1}^{k}(g_i - 1) + 1 + \delta$.

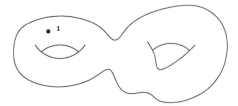

Fig. 3. By smoothing the curve of Fig. 2, we see that its genus is 2

We define the *dual graph* of a pointed nodal curve as follows. It consists of vertices, edges, and "half-edges". The vertices correspond to the irreducible components of the curve, and are labeled with the *geometric* genus of the component. When the genus is 0, the label will be omitted for convenience. The edges correspond to the nodes, and join the corresponding vertices. (Note that an edge can join a vertex to itself.) The half-edges correspond to the labeled points. The dual graph corresponding to Fig. 2 is given in Fig. 4.

Fig. 4. The dual graph to the pointed nodal curve of Fig. 2 (unlabeled vertices are genus 0)

A nodal curve is said to be *stable* if it has finite automorphism group. This is equivalent to a combinatorial condition: (1) each genus 0 vertex of the dual graph has valence at least three, and (2) each genus 1 vertex has valence at least one.

Exercise. Prove this. You may use the fact that a genus $g \geq 2$ curve has finite automorphism group, and that an elliptic curve (i.e. a 1-pointed genus 1 curve) has finite automorphism group. While you are proving this, you may as well show that the automorphism group of a stable genus 0 curve is trivial.

2.5. *Exercise.* Draw all possible stable dual graphs for $g = 0$ and $n \leq 5$; also for $g = 1$ and $n \leq 2$. In particular, show there are no stable dual graphs if $(g, n) = (0,0), (0,1), (0,2), (1,0)$.

Fact. There is a moduli space of stable nodal curves of genus g with n marked points, denoted $\overline{\mathcal{M}}_{g,n}$. There is an open subset corresponding to smooth curves, denoted $\mathcal{M}_{g,n}$. The space $\overline{\mathcal{M}}_{g,n}$ is irreducible, of dimension $3g - 3 + n$, and smooth.

(For Gromov–Witten experts: you can interpret this space as the moduli space of stable maps to a point. But this is in some sense backwards, both historically, and in terms of the importance of both spaces.)

Exercise. Show that $\chi(\mathcal{M}_{g,n}) = (-1)^n \frac{(2g+n-3)!B_{2g}}{2g(2g-2)!}$, using the Harer–Zagier fact earlier (3).

2.6. *Strata.* To each stable graph Γ of genus g with n points, we associate the subset $\mathcal{M}_\Gamma \subset \overline{\mathcal{M}}_{g,n}$ of curves with that dual graph. This translates to the space of curves of a given topological type. Notice that if Γ is the dual graph given in Fig. 4, we can obtain any curve in $\mathcal{M}_\Gamma$ by taking a genus 0 curve with three marked points and gluing two of the points together, and gluing the result to a genus 1 curve with two marked points. (This is most clear in Fig. 2.) Thus each $\mathcal{M}_\Gamma$ is naturally the quotient of a product of $\mathcal{M}_{g',n'}$'s by a finite group. For example, if Γ is as in Fig. 4, $\mathcal{M}_\Gamma = (\mathcal{M}_{0,3} \times \mathcal{M}_{1,2})/S_2$.

These $\mathcal{M}_\Gamma$ give a stratification of $\overline{\mathcal{M}}_{g,n}$, and this stratification is essentially as nice as one could hope. For example, the divisors (the closure of the codimension one strata) meet transversely along smaller strata. The dense open set $\mathcal{M}_{g,n}$ is one stratum; the rest are called *boundary strata*. The codimension 1 strata are called *boundary divisors*.

Notice that even if we were initially interested only in unpointed Riemann surfaces, i.e. in the moduli space $\mathcal{M}_g$, then this compactification forces us to consider $\mathcal{M}_\Gamma$, which in turn forces us to consider pointed nodal curves.

Exercise. By computing $\dim \mathcal{M}_\Gamma$, check that the codimension of the boundary stratum corresponding to a dual graph Γ is precisely the number of edges of the dual graph. (Do this first in some easy case!)

2.7. *Important exercise.* Convince yourself that $\overline{\mathcal{M}}_{0,4} \cong \mathbb{P}^1$. The isomorphism is given as follows. Given four distinct points p_1, p_2, p_3, p_4 on a genus 0 curve (isomorphic to $\mathbb{P}^1$), we may take their cross-ratio $\lambda = (p_4 - p_1)(p_2 - p_3)/(p_4 - p_3)(p_2 - p_1)$, and in turn the cross-ratio determines the points $p_1, \ldots, p_4$ up to automorphisms of $\mathbb{P}^1$. The cross-ratio can take on any value in $\mathbb{P}^1 - \{0, 1, \infty\}$. The three 0-dimensional strata correspond to these three missing points – figure out which stratum corresponds to which of these three points.

Exercise. Write down the strata of $\overline{\mathcal{M}}_{0,5}$, along with which stratum is in the closure of which other stratum (cf. Exercise 2.5).

2.8. Natural Morphisms Among These Moduli Spaces

We next describe some natural maps between these moduli spaces. For example, given any n-pointed genus g curve (where $(g,n) \neq (0,3), (1,1), n > 0$), we can forget the nth point, to obtain an $(n-1)$-pointed nodal curve of genus g. This curve may not be stable, but it can be "stabilized" by contracting all components that are 2-pointed genus 0 curves. This gives us a map $\overline{\mathcal{M}}_{g,n} \to \overline{\mathcal{M}}_{g,n-1}$, which we dub the *forgetful morphism*.

Exercise. Create an example of a dual graph where stabilization is necessary. Also, explain why we excluded the cases $(g,n) = (0,3), (1,1)$.

2.9. *Important exercise.* Interpret $\overline{\mathcal{M}}_{g,n+1} \to \overline{\mathcal{M}}_{g,n}$ as the universal curve over $\overline{\mathcal{M}}_{g,n}$. (This is a bit subtle. Suppose C is a nodal curve, with node p. Which *stable* pointed curve with 1 marked point corresponds to p? Similarly, suppose (C, p) is a pointed curve. Which stable 2-pointed curve corresponds to p?)

Given an $(n_1 + 1)$-pointed curve of genus g_1, and an $(n_2 + 1)$-pointed curve of genus g_2, we can glue the first curve to the second along the last point of each, resulting in an $(n_1 + n_2)$-pointed curve of genus $g_1 + g_2$. This gives a map

$$\overline{\mathcal{M}}_{g_1, n_1 + 1} \times \overline{\mathcal{M}}_{g_2, n_2 + 1} \to \overline{\mathcal{M}}_{g_1 + g_2, n_1 + n_2}.$$

Similarly, we could take a single $(n + 2)$-pointed curve of genus g, and glue its last two points together to get an n-pointed curve of genus $g + 1$; this gives a map

$$\overline{\mathcal{M}}_{g, n+2} \to \overline{\mathcal{M}}_{g+1, n}.$$

We call these last two types of maps *gluing morphisms*.

We call the forgetful and gluing morphisms the *natural morphisms* between moduli spaces of curves.

3 Tautological Cohomology Classes on Moduli Spaces of Curves, and Their Structure

We now define some cohomology classes on these two sorts of moduli spaces of curves, $\mathcal{M}_g$ and $\overline{\mathcal{M}}_{g,n}$. Clearly by Harer and Zagier's Euler-characteristic calculation (3), we should expect some interesting classes, and it is a challenge to name some. Inside the cohomology ring, there is a subring, called the *tautological (sub)ring* of the cohomology ring, that consists informally of the geometrically natural classes. An equally informal definition of the tautological ring is: all the classes you can easily think of. (Of course, this isn't a mathematical statement. But we do not know of a single algebraic class in $H^*(\mathcal{M}_g)$ that can be explicitly written down, that is provably not tautological, even though we expect that they exist.) Hence we care very much about this subring.

The reader may work in cohomology, or in the Chow ring (the algebraic analogue of cohomology). The tautological elements will live naturally in either, and the reader can choose what he or she is most comfortable with. In order to emphasize that one can work algebraically, and also that our dimensions and codimensions are algebraic, I will use the notation of the Chow ring A^i, but most readers will prefer to interpret all statements in the cohomology ring. There is a natural map $A^i \to H^{2i}$, and the reader should be conscious of that doubling of the index.

If α is a 0-cycle on a compact orbifold X, then $\int_X \alpha$ is defined to be its degree.

3.1. Tautological Classes on $\mathcal{M}_g$, Take One

A good way of producing cohomology classes on $\mathcal{M}_g$ is to take Chern classes of some naturally defined vector bundles.

On the universal curve $\pi : \mathcal{C}_g \to \mathcal{M}_g$ over $\mathcal{M}_g$, there is a natural line bundle $\mathcal{L}$; on the fiber C of $\mathcal{C}_g$, it is the line bundle of differentials of C. Define $\psi := c_1(\mathcal{L})$, which lies in $A^1(\mathcal{C}_g)$ (or $H^2(\mathcal{C}_g)$ – but again, we will stick to the language of A^*). Then $\psi^{i+1} \in A^{i+1}(\mathcal{C}_g)$, and as π is a proper map, we can push this class forward to $\mathcal{M}_g$, to get the *Mumford–Morita–Miller κ-class*

$$\boxed{\kappa_i := \pi_* \psi^{i+1}, \quad i = 0, 1, \ldots .}$$

Another natural vector bundle is the following. Each genus g curve (i.e. each point of $\mathcal{M}_g$) has a g-dimensional space of differentials (Sect. 2.1), and the corresponding rank g vector bundle on $\mathcal{M}_g$ is called the *Hodge bundle*, denoted $\mathbb{E}$. (It can also be defined by $\mathbb{E} := \pi_* \mathcal{L}$.) We define the λ-classes by

$$\boxed{\lambda_i := c_i(\mathbb{E}), \quad i = 0, \ldots, g.}$$

We define the *tautological ring* as the subring of the Chow ring generated by the κ-classes. (We will have another definition in Sect. 3.8.) This ring is denoted $R^*(\mathcal{M}_g) \subset A^*(\mathcal{M}_g)$ (or $R^*(\mathcal{M}_g) \subset H^{2*}(\mathcal{M}_g)$).

It is a miraculous "fact" that everything else you can think of seems to lie in this subring. For example, the following generating function identity determines the λ-classes from the κ-classes in an attractive way, and incidentally serves as an advertisement for the fact that generating functions (with coefficients in the Chow ring) are a good way to package information [Fab1, p. 111]:

$$\sum_{i=0}^{\infty} \lambda_i t^i = \exp \left(\sum_{i=1}^{\infty} \frac{B_{2i} \kappa_{2i-1}}{2i(2i-1)} t^{2i-1} \right).$$

3.2. Faber's Conjectures

The study of the tautological ring was begun in Mumford's fundamental paper [Mu], but there was no reason to think that it was particularly well-behaved. But just over a decade ago, Carel Faber proposed a remarkable constellation of conjectures (first in print in [Fab1]), suggesting that the tautological ring has a beautiful combinatorial structure. It is reasonable to state that Faber's conjectures have motivated a great deal of the remarkable progress in understanding the topology of the moduli space of curves over the last decade.

Although Faber's conjectures deal just with the moduli of smooth curves, their creation required knowledge of the compactification, and even of Gromov–Witten theory, as we will later see.

A good portion of Faber's conjectures can be informally summarized as: "$R^*(\mathcal{M}_g)$ behaves like the $((p,p)$-part of the) cohomology ring of a $(g-2)$-dimensional complex projective manifold". We now describe (most of) Faber's conjectures more precisely. I have chosen to cut them into three pieces.

I. "Vanishing/socle" conjecture. $R^i(\mathcal{M}_g) = 0$ for $i > g - 2$, and $R^{g-2}(\mathcal{M}_g) \cong \mathbb{Q}$. This was proved by Looijenga [Lo] and Faber [Fab1, Theorem 2]. (Looijenga's theorem will be stated explicitly below, see Theorem 4.5.) We will prove the "vanishing" part $R^i(\mathcal{M}_g) = 0$ for $i > g - 2$ in Sect. 4.4, and show that $R^{g-2}(\mathcal{M}_g)$ is generated by a single element as a consequence of Theorem 7.10. These statements comprise Looijenga's theorem (Theorem 4.5). The remaining part (that this generator $R^{g-2}(\mathcal{M}_g)$ is non-zero) is a theorem of Faber's, and we omit its proof.

II. Perfect pairing conjecture. The analog of Poincaré duality holds: for $0 \le i \le g - 2$, the natural product $R^i(\mathcal{M}_g) \times R^{g-2-i}(\mathcal{M}_g) \to R^{g-2}(\mathcal{M}_g) \cong \mathbb{Q}$ is a perfect pairing. This conjecture is currently completely open, and is only known in special cases.

We call a ring satisfying **I** and **II** a *Poincaré duality ring of dimension* $g - 2$.

A little thought will convince you that, thanks to **II**, if we knew the "top intersections" (i.e. the products of κ-classes of total degree $g - 2$, as a multiple of the generator of $R^{g-2}(\mathcal{M}_g)$), then we would know the complete structure of the tautological ring. Faber predicts the answer to this as well.

III. Intersection number conjecture (take one). (We will give a better statement in Conjecture 3.23, in terms of a partial compactification of $\mathcal{M}_{g,n}$.) For any n-tuple of non-negative integers $(d_1, \ldots, d_n)$ summing to $g - 2$,

$$
(4) \qquad \frac{(2g - 3 + n)!(2g - 1)!!}{(2g - 1)! \prod_{j=1}^{n}(2d_j + 1)!!} \kappa_{g-2} = \sum_{\sigma \in S_n} \kappa_\sigma
$$

where if $\sigma = (a_{1,1} \cdots a_{1,i_1})(a_{2,1} \cdots a_{2,i_2}) \cdots$ is the cycle decomposition of σ, then κ_σ is defined to be $\prod_j \kappa_{d_{a_{j,1}} + d_{a_{j,2}} + \cdots + d_{a_{j,i_j}}}$. Recall that $(2k - 1)!! = 1 \times 3 \times \cdots \times (2k - 1) = (2k)!/2^k k!$.

For example, we have

$$
\kappa_{i-1}\kappa_{g-i-1} + \kappa_{g-2} = \frac{(2g - 1)!!}{(2i - 1)!!(2g - 2i - 1)!!}\kappa_{g-2}
$$

and

$$
\kappa_1^{g-2} = \frac{1}{g-1} 2^{2g-5}(g-2)!^2 \kappa_{g-2}.
$$

Remarkably, Faber was able to deduce this elegant conjecture from a very limited amount of experimental data.

Faber's intersection number conjecture begs an obvious question: why is this formula so combinatorial? What is the combinatorial structure behind this ring? Faber's alternate description of the intersection number conjecture (Conjecture 3.23) will be even more patently combinatorial.

Faber's intersection number conjecture is now a theorem. Getzler and Pandharipande showed that it is a formal consequence of the *Virasoro conjecture* for the projective plane [GeP]. The Virasoro conjecture is due to the physicists Eguchi, Hori, Xiong, and also the mathematician Sheldon Katz, and deals with the Gromov–Witten invariants of some space X. (See [CK, Sect. 10.1.4] for a statement.) Getzler and Pandharipande show that the Virasoro conjecture in $\mathbb{P}^2$ implies a recursion among the intersection numbers on the (compact) moduli space of stable curves, which in turn is equivalent to a recursion for the top intersections in Faber's conjecture. They then show that the recursions have a unique solution, and that Faber's prediction is a solution.

Givental has announced a proof of the Virasoro conjecture for projective space (and more generally Fano toric varieties) [Giv]. The details of the proof have not appeared, but Y.-P. Lee and Pandharipande are writing a book [LeeP] giving the details. This theorem is really a tour-de-force, and the most important result in Gromov–Witten theory in some time. However, it seems a roundabout and high-powered way of proving Faber's intersection number conjecture. For example, by its nature, it cannot shed light on the combinatorial structure behind the intersection numbers. For this reason, it seems worthwhile giving a more direct argument. At the end of these notes, I will outline a program for tackling this conjecture (joint with the combinatorialists I.P. Goulden and D.M. Jackson), and a proof in a large class of cases.

(There are two other conjectures in this constellation worth mentioning. Faber conjectures that κ_1, ..., $\kappa_{[g/3]}$ generate the tautological ring, with no relations in degrees $\leq [g/3]$. Both Morita [Mo1] and Ionel [I2] have given proofs of the first part of this conjecture a few years ago. Faber also conjectures that $R^*(\mathcal{M}_g)$ satisfies the Hard Lefschetz and Hodge Positivity properties with respect to the class κ_1 [Fab1, Conjecture 1(bis)].

As evidence, Faber has checked that his conjectures hold true in genus up to 21 [Fab4]. I should emphasize that this check is very difficult to do – the rings in question are quite large and complicated! Faber's verification involves some clever constructions, and computer-aided computations.

Morita has recently announced a conjectural form of the tautological ring, based on the representation theory of the symplectic group $Sp(2g, \mathbb{Q})$ [Mo2, Conjecture 1]. This is a new and explicit (and attractive) proposed description of the tautological ring. One might hope that his conjecture may imply Faber's conjecture, and may also be provable.

3.3. Tautological Classes on $\overline{\mathcal{M}}_{g,n}$

We can similarly define a tautological ring on the compact moduli space of stable pointed curves, $\overline{\mathcal{M}}_{g,n}$. In fact here the definition is cleaner, and even sheds new light on the tautological ring of $\mathcal{M}_g$. As before, this ring includes "all classes one can easily think of", and as before, it will be most cleanly described in terms of Chern classes of natural vector bundles. Before we give a formal definition, we begin by discussing some natural classes on $\overline{\mathcal{M}}_{g,n}$.

3.4. *Strata.* We note first that we have some obvious (co)homology classes on $\overline{\mathcal{M}}_{g,n}$, that we didn't have on $\mathcal{M}_g$: the fundamental classes of the (closure of the) strata. We will discuss these classes and their relations at some length before moving on.

In genus 0 (i.e., on $\overline{\mathcal{M}}_{0,n}$), the cohomology (and Chow) ring is generated by these classes. (The reason is that each stratum of the boundary stratification is by (Zariski-)open subsets of affine space.) We will see why the tautological groups are generated by strata in Exercise 4.9.

We thus have generators of the cohomology groups; it remains to find the relations. On $\overline{\mathcal{M}}_{0,4}$, the situation is especially nice. We have checked that $\overline{\mathcal{M}}_{0,4}$ is isomorphic to $\mathbb{P}^1$ (Exercise 2.7), and there are three boundary points. They are homotopic (as any two points on $\mathbb{P}^1$ are homotopic) – and even *rationally equivalent*, the algebraic version of homotopic in the theory of Chow groups.

By pulling back these relations by forgetful morphisms, and pushing forward by gluing morphisms, we get many other relations for various $\overline{\mathcal{M}}_{0,n}$. We dub these *cross-ratio relations*, although they go by many other names in the literature. Keel has shown that these are *all* the relations [Ke].

In genus 1, the tautological ring (although not the cohomology or Chow rings!) are again generated by strata. (We will see why in Exercise 3.28, and again in Exercise 4.9.) We again have cross-ratio relations, induced by a single (algebraic/complex) codimension 1 relation on $\overline{\mathcal{M}}_{0,4}$. Getzler proved a new (codimension 2) relation on $\overline{\mathcal{M}}_{1,4}$ [Ge1, Thm. 1.8] (now known as *Getzler's relation*). (It is remarkable that this relation, on an important compact smooth fourfold, parametrizing four points on elliptic curves, was discovered so late.) Via the natural morphisms, this induces relations on $\overline{\mathcal{M}}_{1,n}$ for all n. Some time ago, Getzler announced that these two sorts of relations were the only relations among the strata [Ge1, par. 2].

In genus 2, there are very natural cohomology classes that are not combinations of strata, so it is now time to describe other tautological classes.

3.5. *Other tautological classes.* Once again, we can define classes as Chern classes of natural vector bundles.

On $\overline{\mathcal{M}}_{g,n}$, for $1 \leq i \leq n$, we define the line bundle $\mathbb{L}_i$ as follows. On the universal curve $\mathcal{C}_{g,n} \to \overline{\mathcal{M}}_{g,n}$, the cotangent space at the fiber above $[(C, p_1, \ldots, p_n)] \in \overline{\mathcal{M}}_{g,n}$ at point p_i is a one-dimensional vector space, and this vector space varies smoothly with $[(C, p_1, \ldots, p_n)]$. This is $\mathbb{L}_i$. More precisely, if $s_i : \overline{\mathcal{M}}_{g,n} \to \mathcal{C}_{g,n}$ is the section of π corresponding to the ith marked point, then $\mathcal{L}_i$ is the pullback by s_i of the sheaf of relative differentials or the relative dualizing sheaf (it doesn't matter which, as the section meets only the smooth locus). Define $\psi_i = c_1(\mathbb{L}_i) \in A^1(\overline{\mathcal{M}}_{g,n})$.

A genus g nodal curve has a g-dimensional vector space of sections of the dualizing line bundle. These vector spaces vary smoothly, yielding the *Hodge bundle* $\mathbb{E}_{g,n}$ on $\overline{\mathcal{M}}_{g,n}$. (More precisely, if π is the universal curve over $\overline{\mathcal{M}}_{g,n}$, and $\mathcal{K}_\pi$ is the relative dualizing line bundle on the universal curve, then $\mathbb{E}_{g,n} := \pi_* \mathcal{K}_\pi$.) Define $\lambda_i := c_i(\mathbb{E}_{g,n})$ on $\overline{\mathcal{M}}_{g,n}$. Clearly the restriction of the Hodge bundle and λ-classes from $\overline{\mathcal{M}}_g$ to $\mathcal{M}_g$ are the same notions defined earlier.

Similarly, there is a more general definition of κ-classes, due to Arbarello and Cornalba [ArbC].

One might reasonably hope that these notions should behave well under the forgetful morphism $\pi : \overline{\mathcal{M}}_{g,n+1} \to \overline{\mathcal{M}}_{g,n}$ (which we can interpret as the universal curve by Exercise 2.9).

Exercise. Show that there is a natural isomorphism $\pi^* \mathbb{E}_{g,n} \cong \mathbb{E}_{g,n+1}$, and hence that $\pi^* \lambda_k = \lambda_k$. (Caution: the two λ_k's in this statement are classes on two different spaces.)

The behavior of the ψ-classes under pullback by the forgetful morphism has a slight twist.

3.6. *Comparison lemma.* $\psi_1 = \pi^* \psi_1 + D_{0,\{1,n+1\}}$.

(Caution: the two ψ_1's in the comparison lemma are classes on two different spaces!) Here $D_{0,\{1,n+1\}}$ means the boundary divisor corresponding to reducible curves with one node, where one component is genus 0 and contains only the marked points p_1 and p_{n+1}. The analogous statement applies with 1 replaced by any number up to n of course.

Exercise (for people with more background). Prove the Comparison Lemma 3.6. (Hint: First show that we have equality away from $D_{0,\{1,n+1\}}$. Hence $\psi_1 = \pi^* \psi_1 + k D_{0,\{1,n+1\}}$ for some integer k, and this integer k can be computed on a single test family.)

As an application:

3.7. *Exercise.* Show that ψ_1 on $\overline{\mathcal{M}}_{0,4}$ is $\mathcal{O}(1)$ (where $\overline{\mathcal{M}}_{0,4} \cong \mathbb{P}^1$, Exercise 2.7).

Exercise. Express ψ_1 explicitly as a sum of boundary divisors on $\overline{\mathcal{M}}_{0,n}$.

We are now ready to define the tautological ring of $\overline{\mathcal{M}}_{g,n}$. We do this by defining the rings for all g and n at once.

3.8. *Definition.* The system of tautological rings $(R^*(\overline{\mathcal{M}}_{g,n}) \subset A^*(\overline{\mathcal{M}}_{g,n}))_{g,n}$ (as g and n vary over all genera and numbers of marked points) is the smallest system of $\mathbb{Q}$-algebras closed under pushforwards by the natural morphisms.

This elegant definition is due to Faber and Pandharipande [FabP3, Sect. 0.1].

Define the tautological ring of any open subset of $\overline{\mathcal{M}}_{g,n}$ by its restriction from $\overline{\mathcal{M}}_{g,n}$. In particular, we can recover our original definition of the tautological ring of $\mathcal{M}_g$ (Sect. 3.1).

It is a surprising fact that everything else you can think of (such as ψ-classes, λ-classes and κ-classes) will lie in this ring. (It is immediate that fundamental classes of strata lie in this ring: they are pushforwards of the fundamental classes of their "component spaces", cf. Sect. 2.6.)

We next give an equivalent description of the tautological *groups*, which will be convenient for many of our arguments, because we do not need to make use of the multiplicative structure. In this description, the ψ-classes play a central role.

3.9. Definition [GrV3, Definition 4.2]. The system of tautological rings $(R^*(\overline{\mathcal{M}}_{g,n}) subset A^*(\overline{\mathcal{M}}_{g,n}))_{g,n}$ is the smallest system of $\mathbb{Q}$- vector spaces closed under pushforwards by the natural morphisms, such that all monomials in $\psi_1, \ldots, \psi_n$ lie in $R^*(\overline{\mathcal{M}}_{g,n})$.

The equivalence of Definition 3.8 and Definition 3.9 is not difficult (see for example [GrV3]).

3.10. Faber-Type Conjectures for $\overline{\mathcal{M}}_{g,n}$, and the Conjecture of Hain–Looijenga–Faber–Pandharipande

In analogy with Faber's conjecture, we have the following.

3.11. Conjecture $R^*(\overline{\mathcal{M}}_{g,n})$ is a Poincaré-duality ring of dimension $3g-3+n$.

This was first asked as a question by Hain and Looijenga [HLo, Question 5.5], first stated as a speculation by Faber and Pandharipande [FabP1, Speculation 3] (in the case $n = 0$), and first stated as a conjecture by Pandharipande [P, Conjecture 1]. In analogy with Faber's conjecture, we break this into two parts.

I. "Socle" conjecture. $R^{3g-3+n}(\overline{\mathcal{M}}_{g,n}) \cong \mathbb{Q}$. This is obvious if we define the tautological ring in terms of cohomology: $H^{2(3g-3+n)}(\overline{\mathcal{M}}_{g,n}) \cong \mathbb{Q}$, and the zero-dimensional strata show that the tautological zero-cycles are not all zero. However, in the tautological Chow ring, the socle conjecture is not at all obvious. Moreover, the conjecture is not true in the full Chow ring – $A_0(\overline{\mathcal{M}}_{1,11})$ is uncountably generated, while the conjecture states that $R_0(\overline{\mathcal{M}}_{1,11})$ has a single generator. (By R_0, we of course mean R^{3g-3+n}.)

We will prove the vanishing conjecture in Sect. 4.6.

II. Perfect pairing conjecture For $0 \leq i \leq 3g-3+n$, the natural product

$$R^i(\overline{\mathcal{M}}_{g,n}) \times R^{3g-3+n-i}(\overline{\mathcal{M}}_{g,n}) \to R^{3g-3+n}(\overline{\mathcal{M}}_{g,n}) \cong \mathbb{Q}$$

is a perfect pairing. (We currently have no idea why this should be true.)

Hence, in analogy with Faber's conjecture, if this conjecture were true, then we could recover the entire ring by knowing the top intersections. This begs the question of how to compute all top intersections.

3.12. Fact/recipe (Mumford and Faber) If we knew the top intersections of ψ-classes, we would know all top intersections. In other words, there is an algorithm to compute all top intersections if we knew the numbers

$$(5) \qquad \int_{\overline{\mathcal{M}}_{g,n}} \psi_1^{a_1} \cdots \psi_n^{a_n}, \qquad \sum a_i = 3g - 3 + n.$$

(This is a worthwhile exercise for people with some familiarity with the moduli space of curves.) This is the basis of Faber's wonderful computer program [Fab2] computing top intersections of various tautological classes. For more information, see [Fab3]. This construction is useful in understanding the definition (Definition 3.9) of the tautological group in terms of the ψ-classes.

Until a key insight of Witten's, there was no a priori reason to expect that these numbers should behave nicely. We will survey three methods of computing these numbers: (1) partial results in low genus; (2) Witten's conjecture; and (3) via the ELSV formula. A fourth (attractive) method was given in Kevin Costello's thesis [C].

3.13. *Top intersections on $\overline{\mathcal{M}}_{g,n}$: partial results in low genus.* Here are two crucial relations among top intersections.

Dilaton equation. If $\overline{\mathcal{M}}_{g,n}$ exists (i.e. there are stable n-pointed genus g curves, or equivalently $2g - 2 + n > 0$), then

$$\int_{\overline{\mathcal{M}}_{g,n+1}} \psi_1^{\beta_1} \psi_2^{\beta_2} \cdots \psi_n^{\beta_n} \psi_{n+1} = (2g - 2 + n) \int_{\overline{\mathcal{M}}_{g,n}} \psi_1^{\beta_1} \cdots \psi_n^{\beta_n}.$$

String equation. If $2g - 2 + n > 0$, then

$$\int_{\overline{\mathcal{M}}_{g,n+1}} \psi_1^{\beta_1} \psi_2^{\beta_2} \cdots \psi_n^{\beta_n} = \sum_{i=1}^{n} \int_{\overline{\mathcal{M}}_{g,n}} \psi_1^{\beta_1} \psi_2^{\beta_2} \cdots \psi_i^{\beta_i - 1} \cdots \psi_n^{\beta_n}$$

(where you ignore terms where you see negative exponents).

Exercise (for those with more experience). Prove these using the Comparison Lemma 3.6.

Equipped with the string equation alone, we can compute all top intersections in genus 0, i.e. $\int_{\overline{\mathcal{M}}_{0,n}} \psi_1^{\beta_1} \cdots \psi_n^{\beta_n}$ where $\sum \beta_i = n - 3$. (In any such expression, some β_i must be 0, so the string equation may be used.) Thus we can recursively solve for these numbers, starting from the base case $\int_{\overline{\mathcal{M}}_{0,3}} \psi_1^0 \psi_2^0 \psi_3^0 = 1$.

Exercise. Show that

$$\int_{\overline{\mathcal{M}}_{0,n}} \psi_1^{a_1} \cdots \psi_n^{a_n} = \binom{n-3}{a_1, \cdots, a_n}.$$

In genus 1, the story is similar. In this case, we need both the string and dilaton equation.

Exercise. Show that any integral

$$\int_{\overline{\mathcal{M}}_{1,n}} \psi_1^{\beta_1} \cdots \psi_n^{\beta_n}$$

can be computed using the string and dilaton equation from the base case $\int_{\overline{\mathcal{M}}_{1,1}} \psi_1 = 1/24$.

We now sketch why the base case $\int_{\overline{\mathcal{M}}_{1,1}} \psi_1 = 1/24$ is true. We calculate this by choosing a finite cover $\mathbb{P}^1 \to \overline{\mathcal{M}}_{1,1}$. Consider a general pencil of cubics in the projective plane. In other words, take two general homogeneous cubic polynomials f and g in three variables, and consider the linear combinations of f and g. The non-zero linear combinations modulo scalars are parametrized by a $\mathbb{P}^1$. Thus we get a family of cubics parametrized by $\mathbb{P}^1$, i.e. $\mathcal{C} \to \mathbb{P}^1$.

You can verify that in this family, there will be twelve singular fibers, that are cubics with one node. One way of verifying this is as follows: $f = g = 0$ consists of nine points $p_1, \ldots, p_9$ (basically by Bezout's theorem – you expect two cubics to meet at nine points). There is a map $\mathbb{P}^2 - \{p_1, \ldots, p_9\} \to \mathbb{P}^1$. If $\mathcal{C}$ is the blow-up of $\mathbb{P}^2$ at the nine points, then this map extends to $\mathcal{C} \to \mathbb{P}^1$, and this is the total space of the family. The (topological) Euler characteristic of $\mathcal{C}$ is the Euler characteristic of $\mathbb{P}^2$ (which is 3) plus 9 (as each blow-up replaces a point by a $\mathbb{P}^1$), i.e. $\chi(\mathcal{C}) = 12$. Considering $\mathcal{C}$ as a fibration over $\mathbb{P}^1$, most fibers are elliptic curves, which have Euler characteristic 0. Hence $\chi(\mathcal{C})$ is the sum of the Euler characteristics of the singular fibers. Each singular fiber is a nodal cubic, which is isomorphic to $\mathbb{P}^1$ with two points glued together (depicted in Fig. 5); this is the union of $\mathbb{C}^*$ (which has Euler characteristic 0) with a point, so $\chi(\mathcal{C})$ is the number of singular fibers. (This argument needs further justification at every point!)

We have a section of $\mathcal{C} \to \mathbb{P}^1$, given by the exceptional fiber E of the blow-up of p_1. Hence we have a moduli map $\mu : \mathbb{P}^1 \to \overline{\mathcal{M}}_{1,1}$ of smooth curves. Clearly it doesn't map $\mathbb{P}^1$ to a point, as some of the fibers are smooth, and twelve are singular. Thus the moduli map μ is surjective (as the image is an irreducible closed set that is not a point). You might suspect that μ has degree 12, as the preimage of the boundary divisor $\Delta \in \overline{\mathcal{M}}_{1,1}$ has 12 preimages, and one can check that μ is nonsingular here. However, we come to one of the twists of stack theory – each point of $\overline{\mathcal{M}}_{1,1}$, including Δ, has degree $1/2$ – each point should be counted with multiplicity one over the size of its automorphism group, and each 1-pointed genus 1 stable curve has precisely one nontrivial automorphism.

Thus $24 \int_{\overline{\mathcal{M}}_{1,1}} \psi_1 = \int_{\mathbb{P}^1} \mu^* \psi_1$, so we wish to show that $\int_{\mathbb{P}^1} \mu^* \psi_1 = 1$. This is an explicit computation on $\mathcal{C} \to \mathbb{P}^1$. You may check that on the blow-up to $\mathcal{C}$, the dualizing sheaf to the fiber at p_1 is given by $-\mathcal{O}(E)|_E$. As $E^2 = -1$, we have $\int_{\mathbb{P}^1} \mu^* \psi_1 = -E^2 = 1$ as desired.

In higher genus, the string and dilaton equation are also very useful.

Exercise. Fix g. Show that using the string and dilaton equation, all of the numbers (5) (for all n) can be computed from a finite number of base cases. The number of base cases required is the number of partitions of $3g - 3$. (It is useful to describe this more precisely, by explicitly describing the generating function for (5) in terms of these base cases.)

3.14. *Witten's conjecture.* So how do we get at these remaining base cases? The answer was given by Witten [W]. (This presentation is not chronological – Witten's conjecture came first, and motivated most of what followed. In particular, it predates Faber's conjectures, and was used to generate the data that led Faber to his conjectures.)

Witten's conjecture (Kontsevich's theorem). Let

$$F_g = \sum_{n \geq 0} \frac{1}{n!} \sum_{k_1, \ldots, k_n} \left(\int_{\overline{\mathcal{M}}_{g,n}} \psi_1^{k_1} \cdots \psi_n^{k_n} \right) t_{k_1} \cdots t_{k_n}$$

be the generating function for the genus g numbers (5), and and let

$$F = \sum F_g \hbar^{2g-2}$$

be the generating function for all genera. (This is *Witten's free energy*, or the *Gromov–Witten potential of a point*.) Then

$$(2n+1) \frac{\partial^3}{\partial t_n \partial t_0^2} F =$$

$$= \left(\frac{\partial^2}{\partial t_{n-1} \partial t_0} F \right) \left(\frac{\partial^3}{\partial t_0^3} F \right) + 2 \left(\frac{\partial^3}{\partial t_{n-1} \partial t_0^2} F \right) \left(\frac{\partial^2}{\partial t_0^2} F \right) + \frac{1}{4} \frac{\partial^5}{\partial t_{n-1} \partial t_0^4} F.$$

Witten's conjecture now has many proofs, by Kontsevich [Ko1], Okounkov–Pandharipande [OP], Mirzakhani [Mi], and Kim–Liu [KiL]. It is a sign of the richness of this conjecture that these proofs are all very different, and all very enlightening in different ways.

The reader should not worry about the details of this formula, and should just look at its shape. Those familiar with integrable systems will recognize this as the Korteweg–de Vries (KdV) equation, in some guise. There was a later reformulation due to Dijkgraaf, Verlinde, and Verlinde [DVV], in terms of the Virasoro algebra. Once again, the reader should not worry about the precise statement, and concentrate on the form of the conjecture. Define differential operators ($n \geq -1$)

$$L_{-1} = -\frac{\partial}{\partial t_0} + \frac{\hbar^{-2}}{2} t_0^2 + \sum_{i=0}^{\infty} t_{i+1} \frac{\partial}{\partial t_i}$$

$$L_0 = -\frac{3}{2} \frac{\partial}{\partial t_1} + \sum_{i=0}^{\infty} \frac{2i+1}{2} t_i \frac{\partial}{\partial t_i} + \frac{1}{16}$$

$$L_n = \sum_{k=0}^{\infty} \frac{\Gamma(m+n+\frac{3}{2})}{\Gamma(k+\frac{1}{2})} (t_k - \delta_{k,1}) \frac{\partial}{\partial t_{n+k}} +$$

$$+ \frac{\hbar^2}{2} \sum_{k=1}^{n-1} (-1)^{k+1} \frac{\Gamma(n-k+\frac{1}{2})}{\Gamma(-k-\frac{1}{2})} \frac{\partial}{\partial t_k} \frac{\partial}{\partial t_{n-k-1}} \qquad (n > 0)$$

These operators satisfy $[L_m, L_n] = (m-n) L_{m+n}$.

Exercise. Show that $L_{-1} e^F = 0$ is equivalent to the string equation. Show that $L_0 e^F = 0$ is equivalent to the dilaton equation.

Witten's conjecture is equivalent to $L_n e^F = 0$ for all n. These equations let you inductively solve for the coefficients of F, and hence compute all these numbers.

3.15. *The Virasoro conjecture.* The Virasoro formulation of Witten's conjecture has a far-reaching generalization, the *Virasoro conjecture* described earlier. Instead of top intersections on the moduli space of curves, it addresses top (virtual) intersections on the moduli space of maps of curves to some space X. Givental's proof (to be explicated by Lee and Pandharipande) for the case of projective space (and more generally Fano toric varieties) was mentioned earlier. It is also worth mentioning Okounkov and Pandharipande's proof in the case where X is a curve; this is also a tour-de-force.

3.16. *Hurwitz numbers and the ELSV formula.* We can also recover these top intersections via the old-fashioned theme of branched covers of the projective line, the very technique that let us compute the dimension of the moduli space of curves, and of the Picard variety Sect. 2.2.

Fix a genus g, a degree d, and a partition of d into n parts, $\alpha_1 + \cdots + \alpha_n = d$, which we write as $\alpha \vdash d$. Let

$$(6) \qquad\qquad r := 2g + d + n - 2.$$

Fix $r + 1$ points $p_1, \ldots, p_r, \infty \in \mathbb{P}^1$. Define the *Hurwitz number* H_α^g to be the number of branched covers of $\mathbb{P}^1$ by a Riemann surface, that are unbranched away from $p_1, \ldots, p_r, \infty$, such that the branching over ∞ is given by $\alpha_1, \ldots, \alpha_n$ (i.e. there are n preimages of ∞, and the branching at the ith preimage is of order α_i, i.e. the map is analytically locally given by $t \mapsto t^{\alpha_i}$), and there is the simplest possible branching over each p_i, i.e. the branching is given by $2 + 1 + \cdots + 1 = d$. (To describe this *simple branching* more explicitly: above any such branch point, $d - 2$ of the sheets are unbranched, and the remaining two sheets come together. The analytic picture of the two sheets is the projection of the parabola $y^2 = x$ to the x-axis in $\mathbb{C}^2$.) We consider the n preimages of ∞ to be labeled. Caution: in the literature, sometimes the preimages of ∞ are *not labeled*; that definition of Hurwitz number will be smaller than ours by a factor of $\#\mathrm{Aut}\alpha$, where $\mathrm{Aut}\alpha$ is the subgroup of S_n fixing the n-tuple $(\alpha_1, \ldots, \alpha_n)$ (e.g. if $\alpha = (2, 2, 2, 5, 5)$, then $\#\mathrm{Aut}\alpha = 3!2!$).

One technical point: each cover is counted with multiplicity 1 over the size of the automorphism group of the cover.

Exercise. Use the Riemann–Hurwitz formula (2) to show that if the cover is connected, then it has genus g.

Experts will recognize these as relative descendant Gromov–Witten invariants of $\mathbb{P}^1$; we will discuss relative Gromov–Witten invariants of $\mathbb{P}^1$ in Sect. 5. However, they are something much more down-to-earth. The following result shows that this number is a purely combinatorial object. In particular, there are a finite number of such covers.

3.17. *Proposition.*

$$H_\alpha^g = \#\left\{ (\sigma_1, \ldots, \sigma_r) : \sigma_i \text{ transpositions generating } S_d, \prod_{i=1}^{r} \sigma_i \in C(\alpha) \right\}$$

$$\#\mathrm{Aut}\alpha/d!,$$

where the σ_i are transpositions generating the symmetric group S_d, and $\mathcal{C}(\alpha)$ is the conjugacy class in S_d corresponding to partition α.

Before we give the proof, we make some preliminary comments. As an example, consider $d = 2$, $\alpha = 2$, g arbitrary, so $r = 2g + 1$. The above formula gives $H^g_\alpha = 1/2$, which at first blush seems like nonsense – how can we count covers and get a non-integer? Remember however the combinatorial/stack-theoretic principal that objects should be counted with multiplicity 1 over the size of their automorphism group. Any double cover of this sort always has a non-trivial involution (the "hyperelliptic involution"). Hence there is indeed one cover, but it is counted as "half a cover". Fortunately, this is the only case of Hurwitz numbers for which this is an issue. The reader may want to follow this particular case through in the proof.

Proof of Proposition 3.17. Pick another point $0 \in \mathbb{P}^1$ distinct from $p_1, \ldots, p_r, \infty$. Choose branch cuts from 0 to $p_1, \ldots, p_r, \infty$ (non-intersecting paths from 0 to p_1, 0 to p_2, ..., 0 to ∞) such that their cyclic order around 0 is $p_1, \ldots, p_r, \infty$. Suppose $C \to \mathbb{P}^1$ is one of the branched covers counted by H^g_α. Then label the d preimages of 0 with 1 through d in some way. We will count these labeled covers, and divide by $d!$ at the end. Now cut along the preimages of the branch-cuts. As $\mathbb{P}^1$ minus the branch-cuts is homeomorphic to a disc, which is simply connected, its preimage must be d copies of the disc, labelled 1 through d according to the label on the preimage of 0. We may reconstruct $C \to \mathbb{P}^1$ by determining how to glue these sheets together along the branch cuts. The monodromy of the cover $C \to \mathbb{P}^1$ around p_i is an element σ_i of S_d, and this element will be a transposition, corresponding to the two sheets being interchanged above that branch point. Similarly, the monodromy around ∞ is also an element σ_∞ of S_d, with cycle type α. The cover has the additional data of the bijection of the cycles with the parts of α. In $\pi_1(\mathbb{P}^1 - \{p_1, \ldots, p_r, \infty\})$, the loops around p_1, ..., p_r, ∞ multiply to the identity, so $\sigma_1\sigma_2\cdots\sigma_r\sigma_\infty = e$. (Here we use the fact that the branch cuts meet 0 in this particular order.) Thus $\sigma_\infty^{-1} = \sigma_1\cdots\sigma_r$. This is the only relation among these generators of $\pi_1(\mathbb{P}^1 - \{p_1, \ldots, p_r, \infty\})$. Furthermore, the cover C is connected, meaning that we can travel from any one of the d sheets to any of the others, necessarily by travelling around the branch points. This implies that the σ_1, ..., σ_r, σ_∞ (and hence just the σ_1, ..., σ_r) generate a transitive subgroup of S_d. But the only transitive subgroup of S_d containing a transposition σ_1 is all of S_d.

Conversely, given the data of transposition σ_1, ..., σ_r generating S_d, with product of cycle type α, along with a labelling of the parts of the product (of which there are $\#\mathrm{Aut}\alpha$), we can construct a connected cover $C \to \mathbb{P}^1$, by the Riemann existence theorem. Thus, upon forgetting the labels of the d sheets, we obtain the desired equality. $\qquad\square$

The above proof clearly extends to deal with more general Hurwitz numbers, where arbitrary branching is specified over each of a number of points.

Proposition 3.17 shows that any Hurwitz number may be readily computed by hand or by computer. What is interesting is the structure behind them. In 1891, Hurwitz [H] showed that

$$\text{(7)} \qquad H_\alpha^0 = r! d^{n-3} \prod \left(\frac{\alpha_i^{\alpha_i}}{\alpha_i!} \right).$$

By modern standards, he provided an outline of a proof. His work was forgotten by a large portion of the mathematics community, and later people proved special cases, including Dénes [D] in the case $n = 1$, Arnol'd [Arn] in the case $n = 2$. In the case $n = d$ (so $\alpha = 1^d$) it was stated by the physicists Crescimanno and Taylor [CT], who apparently asked the combinatorialist Richard Stanley about it, who in turn asked Goulden and Jackson. Goulden and Jackson independently discovered and proved Hurwitz' original theorem in the mid-nineties [GJ1]. Since then, many proofs have been given, including one by myself using moduli of curves [V1].

Goulden and Jackson studied the problem for higher genus, and conjectured a structural formula for Hurwitz numbers in general. Their polynomiality conjecture [GJ2, Conjecture 1.2] implies the following.

3.18. *Goulden–Jackson polynomiality conjecture (one version).* For each g, n, there is a symmetric polynomial $P_{g,n}$ in n variables, with monomials of homogeneous degree between $2g - 3 + n$ and $3g - 3 + n$, such that

$$H_\alpha^g = r! \prod_{i=1}^n \left(\frac{\alpha_i^{\alpha_i}}{\alpha_i!} \right) P_{g,n}(\alpha_1, \dots, \alpha_n).$$

The reason this conjecture (and the original version) is true is an amazing theorem of Ekedahl, Lando, M. Shapiro, and Vainshtein.

3.19. *Theorem (ELSV formula, by Ekedahl, Lando, M. Shapiro, and Vainshtein [ELSV1, ELSV2]).*

$$\text{(8)} \qquad H_\alpha^g = r! \prod_{i=1}^n \left(\frac{\alpha_i^{\alpha_i}}{\alpha_i!} \right) \int_{\overline{\mathcal{M}}_{g,n}} \frac{1 - \lambda_1 + \dots + (-1)^g \lambda_g}{(1 - \alpha_1 \psi_1) \cdots (1 - \alpha_n \psi_n)}$$

(if $\overline{\mathcal{M}}_{g,n}$ exists).

We will give a proof in Sect. 6.1.

Here is how to interpret the right side of the equation. Note that the α_i are integers, and the ψ_i's and λ_k's are cohomology (or Chow) classes. Formally invert the denominator, e.g.

$$\frac{1}{1 - \alpha_1 \psi_1} = 1 + \alpha_1 \psi_1 + \alpha_1^2 \psi_1^2 + \cdots.$$

Then multiply everything out inside the integral sign, and discard all but the summands of total codimension $3g - 3 + n$ (i.e. dimension 0). Then take the degree of this cohomology class.

For example, if $g = 0$ and $n = 4$, we get

$$H_\alpha^g = r! \prod_{i=1}^{4} \left(\frac{\alpha_i^{\alpha_i}}{\alpha_i!} \right) \int_{\overline{\mathcal{M}}_{0,4}} \frac{1 - \lambda_1 + \cdots \pm \lambda_g}{(1 - \alpha_1 \psi_1) \cdots (1 - \alpha_4 \psi_4)}$$

$$= r! \prod_{i=1}^{4} \left(\frac{\alpha_i^{\alpha_i}}{\alpha_i!} \right) \int_{\overline{\mathcal{M}}_{0,4}} (1 + \alpha_1 \psi_1 + \cdots) \cdots (1 + \alpha_4 \psi_4 + \cdots)$$

$$= r! \prod_{i=1}^{4} \left(\frac{\alpha_i^{\alpha_i}}{\alpha_i!} \right) \int_{\overline{\mathcal{M}}_{0,4}} (\alpha_1 \psi_1 + \cdots + \alpha_4 \psi_4)$$

$$= r! \prod_{i=1}^{4} \left(\frac{\alpha_i^{\alpha_i}}{\alpha_i!} \right) (\alpha_1 + \cdots + \alpha_4) \qquad \text{(Exercise 3.7)}$$

$$= r! \prod_{i=1}^{4} \left(\frac{\alpha_i^{\alpha_i}}{\alpha_i!} \right) d.$$

Exercise. Recover Hurwitz' original formula (7) from the ELSV-formula, at least if $n \geq 3$.

More generally, expanding the integrand of (8) yields

$$(9) \qquad \sum_{a_1 + \cdots + a_n + k = 3g-3+n} \left((-1)^k \left(\int_{\overline{\mathcal{M}}_{g,n}} \psi_1^{a_1} \cdots \psi_n^{a_n} \lambda_k \right) \right) (\alpha_1^{a_1} \cdots \alpha_n^{a_n}).$$

This is a polynomial in $\alpha_1, \ldots, \alpha_n$ of homogeneous degree between $2g - 3 + n$ and $3g - 3 + n$. Thus this explains the mystery polynomial in the Goulden–Jackson Polynomiality Conjecture 3.18 – and the coefficients turn out to be top intersections on the moduli space of curves! (The original polynomiality conjecture was actually different, and some translation is necessary in order to make the connection with the ELSV formula [GJV1].)

There are many other consequences of the ELSV formula; see [ELSV2, GJV1] for surveys.

We should take a step back to see how remarkable the ELSV formula is. To any reasonable mathematician, Hurwitz numbers (as defined by Proposition 3.17) are purely discrete, combinatorial objects. Yet their structure is fundamentally determined by the topology of the moduli space of curves. Put more strikingly – the combinatorics of transpositions in the symmetric group leads inexorably to the tautological ring of the moduli space of curves!

3.20. We return to our original motivation for discussing the ELSV formula: computing top intersections of ψ-classes on the moduli space of curves $\overline{\mathcal{M}}_{g,n}$. Fix g and n. As stated earlier, any given Hurwitz number may be readily computed (and this can be formalized elegantly in the language of generating functions). Thus any number of values of $P_{g,n}(\alpha_1, \ldots, \alpha_n)$ may be computed. However, we know that $P_{g,n}$ is a symmetric polynomial of known degree, and it is straightforward to show that one can determine the coefficients of a polynomial of known degree from enough values. In particular, from (9), the coefficients of the highest-degree terms in $P_{g,n}$ are precisely the top intersections of ψ-classes.

This is a powerful perspective. As an important example, Okounkov and Pandharipande used the ELSV formula to prove Witten's conjecture.

3.21. Back to Faber-Type Conjectures

This concludes our discussion of Faber-type conjectures for $\overline{\mathcal{M}}_{g,n}$. I have two more remarks about Faber-type conjectures. The first is important, the second a side-remark.

3.22. *Faber's intersection number conjecture on $\mathcal{M}_g$, take two.* We define the moduli space of n-pointed genus g curves with "rational tails", denoted $\mathcal{M}_{g,n}^{rt}$, as follows. We define $\mathcal{M}_{g,n}^{rt}$ as the dense open subset of $\overline{\mathcal{M}}_{g,n}$ parametrizing pointed nodal curves where one component is nonsingular of genus g (and the remaining components form trees of genus 0 curves sprouting from it – hence the phrase "rational tails"). If $g > 1$, then $\mathcal{M}_{g,n}^{rt} = \pi^{-1}(\mathcal{M}_g)$, where $\pi : \overline{\mathcal{M}}_{g,n} \to \overline{\mathcal{M}}_g$ is the forgetful morphism. Note that $\mathcal{M}_g^{rt} = \mathcal{M}_g$.

We may restate Faber's intersection number conjecture (for $\mathcal{M}_g$) in terms of this moduli space. By our redefinition of the tautological ring on $\mathcal{M}_g$ in Sect. 3.3 (Definition 3.9, using also Faber's constructions of Sect. 3.12), the "top intersections" are determined by $\pi_* \psi_1^{a_1} \cdots \psi_n^{a_n}$ (where $\pi : \mathcal{M}_{g,n}^{rt} \to \mathcal{M}_g$) for $\sum a_i = g - 2 + n$.

Then Faber's intersection number conjecture translates to the following.

3.23. *Faber's intersection number conjecture (take two). If all $a_i > 1$, then*

$$\psi_1^{a_1} \cdots \psi_n^{a_n} = \frac{(2g - 3 + n)!(2g - 1)!!}{(2g - 1)! \prod_{j=1}^n (2a_j - 1)!!} [generator] \quad for \sum a_i = g - 2 + n$$

where $[generator] = \kappa_{g-2} = \pi_* \psi_1^{g-1}$.

(This reformulation is also due to Faber.) This description is certainly more beautiful than the original one (4), which suggests that we are closer to the reason for it to be true.

3.24. The other conjectures of Faber were extended to $\mathcal{M}_{g,n}^{rt}$ by Pandharipande [P, Conjecture 1].

3.25. *Remark: Faber-type conjectures for curves of compact type.* Based on the cases of $\mathcal{M}_g$ and $\mathcal{M}_{g,n}$, Faber and Pandharipande made another conjecture for curves of "compact type". A curve is said to be of *compact type* if its Jacobian is compact, or equivalently if its dual graph has no loops, or equivalently, if the curve has no nondisconnecting nodes. Define $\mathcal{M}_{g,n}^c \subset \overline{\mathcal{M}}_{g,n}$ to be the moduli space of curves of compact type. It is $\overline{\mathcal{M}}_{g,n}$ minus an irreducible divisor, corresponding to singular curves with one irreducible component (called Δ_0, although we will not use this notation).

3.26. *Conjecture (Faber–Pandharipande [FabP1, Spec. 2], [P, Conjecture 1].*
$R^*(\mathcal{M}_g^c)$ *is a Poincaré duality ring of dimension* $2g - 3$.

Again, this has a *vanishing/socle* part and a *perfect pairing* part. There is something that can be considered the corresponding *intersection number* part, Pandharipande and Faber's λ_g theorem [FabP2].

We will later (Sect. 4.7) give a proof of the vanishing/socle portion of the conjecture, that $R^i(\mathcal{M}_g^c) = 0$ for $i > 2g - 3$, and is 1-dimensional if $i = 2g - 3$. The perfect pairing part is essentially completely open.

3.27. Other Relations in the Tautological Ring

We have been concentrating on top intersections in the tautological ring. I wish to discuss more about other relations (in smaller codimension) in the tautological ring.

In genus 0, as stated earlier (Sect. 3.4), all classes on $\overline{\mathcal{M}}_{g,n}$ are generated by the strata, and the only relations among them are the cross-ratio relations. We have also determined the ψ-classes in terms of the boundary classes.

In genus 1, we can verify that ψ_1 can be expressible in terms of boundary strata. On $\overline{\mathcal{M}}_{1,1}$, if the boundary point is denoted δ_0 (the class of the nodal elliptic curve shown in Fig. 5), we have shown $\psi_1 = \delta_0/12$. (Reason: we proved it was true on a finite cover, in the course of showing that $\int_{\overline{\mathcal{M}}_{1,1}} \psi_1 = 1/24$.) We know how to pull back ψ-classes by forgetful morphisms, so we can now verify the following.

Exercise. Show that in the cohomology group of $\overline{\mathcal{M}}_{1,n}$, ψ_i is equivalent to a linear combination of boundary divisors. (Hint: use the Comparison Lemma 3.6.)

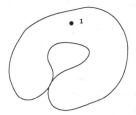

Fig. 5. The curve corresponding to the point $\delta_0 \in \overline{\mathcal{M}}_{1,1}$

3.28. *Slightly trickier exercise.* Use the above to show that the tautological ring in genus 1 is generated (as a group) by boundary classes. (This fact was promised in Sect. 3.4.)

In genus 2, this is no longer true: ψ_1 is not equivalent to a linear combination of boundary strata on $\overline{\mathcal{M}}_{2,1}$. However, in 1983, Mumford showed that ψ_1^2 (on $\overline{\mathcal{M}}_{2,1}$) is a combination of boundary strata ([Mu], see also [Ge2, Equation (4)]); in 1998, Getzler showed the same for $\psi_1\psi_2$ (on $\overline{\mathcal{M}}_{2,2}$) [Ge2]. These two results can be used to show that on $\overline{\mathcal{M}}_{2,n}$, all tautological classes are linear combinations of strata, and classes "constructed using ψ_1 on $\overline{\mathcal{M}}_{2,1}$".

Figure 6 may help elucidate what classes we mean – they correspond to dual graphs, with at most one marking ψ on an edge incident to one genus 2 component. The class in question is defined by gluing together the class of ψ_i on $\overline{\mathcal{M}}_{2,v}$ corresponding to that genus 2 component (where v is the valence, and i corresponds to the edge labeled by ψ) with the fundamental classes of the $\overline{\mathcal{M}}_{0,v_j}$'s corresponding to the other vertices. The question then arises: what are the relations among these classes? On top of the cross-ratio and Getzler relation, there is a new relation due to Belorousski and Pandharipande, in codimension 2 on $\overline{\mathcal{M}}_{2,3}$ [BP]. We do not know if these three relations generate all the relations. (All the genus 2 relations mentioned in this paragraph are given by explicit formulas, although they are not pretty to look at.)

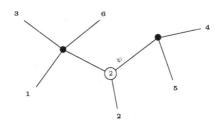

Fig. 6. A class on $\overline{\mathcal{M}}_{2,6}$ – a codimension 1 class on a boundary stratum, constructed using ψ_1 on $\overline{\mathcal{M}}_{2,3}$ and gluing morphisms

In general genus, the situation should get asymptotically worse as g grows. However, there is a general statement that can be made:

3.29. *Getzler's conjecture [Ge2, footnote 1] (Ionel's theorem [I1]).* If $g > 0$, all degree g polynomials in ψ-classes vanish on $\mathcal{M}_{g,n}$ (hence live on the boundary of $\overline{\mathcal{M}}_{g,n}$).

We will interpret this result as a special case of a more general result (Theorem $\star$), in Sect. 4.3. In keeping with the theme of this article, the proof will be Gromov–Witten theoretic.

3.30. Y.-P. Lee's Invariance Conjecture There is another general statement that may well give *all* the relations in *every* genus: Y.-P. Lee's Invariance conjecture. It is certainly currently beyond our ability to either prove or disprove it, although the first part is already a theorem (Theorem 3.31). Lee's conjecture is strongly motivated by Gromov–Witten theory.

Before we state the conjecture, we discuss the consequences and evidence. All of the known relations in the tautological rings are consequences of the conjecture. For example, the genus 2 implications are shown by Arcara and Lee in [ArcL1]. They then predicted a *new* relation in $\overline{\mathcal{M}}_{3,1}$ in [ArcL2]. Simultaneously and independently, this relation was proved by Kimura and X. Liu [KL]. This seems to be good evidence for the conjecture being true.

More recently, the methods behind the conjecture have allowed Lee to turn these predictions into proofs, *not conditional on the truth of the conjecture* [Lee2]. Thus for example Arcara and Lee's work yields a proof of the new relation on $\overline{\mathcal{M}}_{3,1}$.

We now give the statement. The conjecture is most naturally expressed in terms of the tautological rings of *possibly-disconnected* curves. The definition of a stable possibly-disconnected curve is the same as that of a stable curve, except the curve is not required to be connected. We denote the moduli space of n-pointed genus g possibly-disconnected curves by $\overline{\mathcal{M}}_{g,n}^{\bullet}$. The reader can quickly verify that our discussion of the moduli space of curves carries over without change if we consider possibly-disconnected curves. For example, $\overline{\mathcal{M}}_{g,n}^{\bullet}$ is nonsingular and pure-dimensional of dimension $3g - 3 + n$ (although not in general irreducible). It contains $\overline{\mathcal{M}}_{g,n}$ as a component, so any statements about $\overline{\mathcal{M}}_{g,n}^{\bullet}$ will imply statements about $\overline{\mathcal{M}}_{g,n}$. Note that the disjoint union of two curves of arithmetic genus g and h is a curve of arithmetic genus $g + h - 1$: Euler characteristics add under disjoint unions. Note also that a possibly-disconnected marked curve is stable if and only if all of its connected components are stable.

Exercise. Show that $\overline{\mathcal{M}}_{-1,6}^{\bullet}$ is a union of $\binom{6}{3}/2$ points – any 6-pointed genus -1 stable curve must be the disjoint union of two $\mathbb{P}^1$'s, with 3 of the 6 labeled points on each component.

Exercise. Show that any component of $\overline{\mathcal{M}}_{g,n}^{\bullet}$ is the quotient of a product of $\overline{\mathcal{M}}_{g',n'}$'s by a finite group.

Tautological classes are generated by classes corresponding to a dual graph, with each vertex (of genus g and valence n, say) labeled by some cohomology class on $\overline{\mathcal{M}}_{g,n}^{\bullet}$ (possibly the fundamental class); call this a decorated dual graph. (We saw an example of a decorated dual graph in Fig. 6. Note that ψ-classes will always be associated to some half edge.) Decorated dual graphs are not required to be connected. If Γ is a decorated dual graph (of genus g with n tails, say), let $\dim \Gamma$ be the dimension of the corresponding class in $A_*(\overline{\mathcal{M}}_{g,n}^{\bullet})$.

For each positive integer l, we will describe a linear operator $\mathfrak{r}_l$ that sends formal linear combinations of decorated dual graphs to formal linear combinations of decorated dual graphs. It is homogeneous of degree $-l$: it sends (dual graphs corresponding to) dimension k classes to (dual graphs corresponding to) dimension $k - l$ classes.

We now describe its action on a single decorated dual graph Γ of genus g with n marked points (or half-edges), labeled 1 through n. Then $\mathfrak{r}_l(\Gamma)$ will be a formal linear combination of other graphs, each of genus $g - 1$ with $n + 2$ marked points.

There are three types of contributions to $\mathfrak{r}_l(\Gamma)$. (In each case, we discard any graph that is not stable.)

1. *Edge-cutting.* There are two contributions for each *directed* edge, i.e. an edge with chosen starting and ending point. (Caution: there are two possible

directions for each edge in general, except for those edges that are "loops", connecting a single vertex to itself. In this case, both directions are considered the same.) We cut the edge, regarding the two half-edges as "tails", or marked points. The starting half-edge is labeled $n + 1$, and the ending half-edge is labeled $n + 2$. One summand will correspond to adding an extra decoration of ψ^l to point $n + 1$. (In other words, ψ^l_{n+1} is multiplied by whatever cohomology class is already decorating that vertex.) A second summand will correspond to the adding an extra decoration of ψ^l to point $n + 2$, and this summand appears with multiplicity $(-1)^{l-1}$.

2. Genus reduction For each vertex we produce l graphs as follows. We reduce the genus of the vertex by 1, and add two new tails to this vertex, labelled $n + 1$ and $n + 2$; we decorate them with ψ^m and ψ^{l-1-m} respectively, where $0 \leq m \leq l - 1$. Each such graph is taken with multiplicity $(-1)^{m+1}$.

3. Vertex-splitting. For each vertex, we produce a number of graphs as follows. We split the vertex into two. The first new vertex is given the tail $n + 1$, and the second is given the tail $n + 2$. The two new tails are decorated by ψ^m and ψ^{l-1-m} respectively, where $0 \leq m \leq l - 1$. We then take one such graph for each choice of splitting of the genus $g = g_1 + g_2$ and partitioning of the other incident edges. Each such graph is taken with multiplicity $(-1)^{m+1}$.

Then $\mathfrak{r}_l(\Gamma)$ is the sum of the above summands. Observe that when l is odd (resp. even), the result is symmetric (resp. anti-symmetric) in the labels $n + 1$ and $n + 2$.

By linearity, this defines the action of $\mathfrak{r}_l$ on any linear combination of directed graphs.

3.31. Y.P. Lee's invariance theorem [Lee2]. *If $\sum c_i \Gamma_i = 0$ holds in $A^*(\overline{\mathcal{M}}^\bullet_{g,n})$, then $\mathfrak{r}_l(\sum c_i \Gamma_i) = 0$ in $A^*(\overline{\mathcal{M}}_{g-1,n+2})$.*

This was invariance conjecture 1 of [Lee1]. It gives a necessary condition for a tautological class to be zero. Hence for example it can be used to determine the coefficients of a tautological equation, if we already know there is one by other means. The theorem also implies that $\mathfrak{r}_l$ is well-defined at the level of tautological rings (i.e. compatible with descending). The converse of Theorem 3.31 is also conjectured to be true, and would be a sufficient condition for a candidate tautological equation to hold true:

3.32. Y.-P. Lee's invariance conjecture [Lee1, Conjecture 2]. *If $\sum c_i \Gamma_i$ has* **positive** *pure dimension, and $\mathfrak{r}_l(\sum c_i \Gamma_i) = 0$ in $A^*(\overline{\mathcal{M}}_{g-1,n+2})$, then $\sum c_i \Gamma_i = 0$ holds in $A^*(\overline{\mathcal{M}}^\bullet_{g,n})$.*

Theorem 3.31 and Conjecture 3.32 can be used to produce tautological equations inductively! The base case is when $\dim \overline{\mathcal{M}}^\bullet_{g,n} = 0$, which is known: we will soon show (Sect. 4.6) that $R_0(\overline{\mathcal{M}}_{g,n}) \cong \mathbb{Q}$. (We write $R^{3g-3+n}(\overline{\mathcal{M}}_{g,n})$ as $R_0(\overline{\mathcal{M}}_{g,n})$ to remind the reader that the statement is about tautological 0-cycles.) Hence dimension 0 tautological classes on $\overline{\mathcal{M}}_{g,n}$ are determined by their degree (and dimension 0 tautological classes on $\overline{\mathcal{M}}^\bullet_{g,n}$ are determined by their degree on each connected component). Note that the algorithm is a finite process: the dimension l relations on $\overline{\mathcal{M}}_{g,n}$ or $\overline{\mathcal{M}}_{g,n}^{\bullet}$ produced by this algorithm are produced after a finite number of steps.

Even more remarkably, this seems to produce *all* tautological relations:

3.33. *Y.-P. Lee's invariance conjecture, continued [Lee1, Conjecture 3].* Conjecture 3.32 will produce **all** tautological equations inductively.

A couple of remarks are in order. Clearly this is a very combinatorial description. It was dictated by Gromov–Witten theory, as explained in [Lee1]. In particular, it uses the fact that all tautological equations are invariant under the action of lower triangular subgroups of the twisted loop groups, and proposes that they are the only equations invariant in this way.

In order to see the magic of this conjecture in action, and to get experience with the $\mathfrak{r}_l$ operators, it is best to work out an example. The simplest dimension 1 relation is the following.

Exercise. Show that the pullback of the (dimension 0) cross-ratio relation (Sect. 3.4) on $\overline{\mathcal{M}}_{0,4}$ to a (dimension 1) relation on $\overline{\mathcal{M}}_{0,5}$ is implied by the Invariance Conjecture. (Some rather beautiful cancellation happens.)

3.34. *Final remarks on relations in the tautological ring.* This continues to be an active area of research. We point out for example Arcara and Sato's recent article [ArcS] using localization in Gromov–Witten theory to compute the class $\sum_{k=0}^{g}(-1)^{g-k}\psi_1^k\lambda_{g-k}$ explicitly as a sum of boundary classes on $\overline{\mathcal{M}}_{g,1}$.

4 A Blunt Tool: Theorem ⋆ and Consequences

We now describe a blunt tool from which much of the previously described structure of the tautological ring follows. Although it is statement purely about the stratification of the moduli space of curves, we will see (Sect. 6.3) that it is proved via Gromov–Witten theory.

4.1. *Theorem ⋆ [GrV3]. Any tautological class of codimension i is trivial away from strata satisfying*

$$\boxed{\# \text{ genus } 0 \text{ vertices } \geq i - g + 1.}$$

(Recall that the genus 0 vertices correspond to components of the curve with geometric genus 0.)

More precisely, any tautological class is zero upon restriction to the (large) open set corresponding to the open set corresponding to

$$\# \text{ genus } 0 \text{ vertices } < i - g + 1.$$

Put another way: given any tautological class of codimension i, you can move it into the set of curves with at least $i - g + 1$ genus 0 components. A third formulation is that the tautological classes of codimension i are pushed forward from classes on the locus of curves with at least $i - g + 1$ genus 0 components.

We remark that this is false for the Chow ring as a whole – this is fundamentally a statement about tautological classes.

We will discuss the proof in Sect. 6.3, but first we give consequences. There are in some sense four morals of this result.

First, this is the fundamental geometry behind many of the theorems we have been discussing. We will see that they follow from Theorem ⋆ by straightforward combinatorics. As a sign of this, we will often get strengthenings of what was known or conjectured previously.

Second, this suggests the potential importance of a filtration of the moduli space by the number of genus 0 components. It would be interesting to see if this filtration really is fundamental, for example if it ends up being relevant in understanding the moduli space of curves in another way. So far this has not been the case.

Third, as we will see from the proof, once one knows a clean statement of what one wants to prove, the proof is relatively straightforward, at least in outline.

And fourth, the proof will once again show the centrality of Gromov–Witten theory to the study of the moduli of curves.

4.2. Consequences of Theorem ⋆

We begin with a warm-up example.

4.3. Theorem ⋆ implies Getzler's conjecture 3.29 (Ionel's theorem). Any degree g monomial is a codimension g tautological class, which vanishes on the open set of $\overline{\mathcal{M}}_{g,n}$ corresponding to curves with no genus 0 components. If $g > 0$, this is non-empty and includes $\mathcal{M}_{g,n}$.

In particular: (1) we get a proof of Getzler's conjecture; (2) we see that more classes vanish on this set – all tautological classes of degree at least g, not just polynomials in the ψ-classes; (3) we observe that the classes vanish on a bigger set than $\mathcal{M}_{g,n}$, and that what is relevant is not the smoothness of the curves, but the fact that they have no genus 0 components. (4) This gives a moral reason for Getzler's conjecture not to hold in genus 0.

4.4. Theorem ⋆ implies the first part of Looijenga's Theorem (Faber's vanishing conjecture). Recall (Sect. 3.2) that Looijenga's Theorem is part of the "vanishing" part of Faber's conjectures:

4.5. Theorem [Lo]. *We have* $R^i(\mathcal{M}_g) = 0$ *for* $i > g - 2$, $\dim R^{g-2}(\mathcal{M}_g) \leq 1$.

We will show that Theorem ⋆ implies the first part now; we will show the second part as a consequence of Theorem 7.10.

First, if the codimension of a tautological class is greater than or equal to g, then we get vanishing on the open set where there are no genus 0 components, so we get vanishing for the same reason as in Getzler's conjecture.

The case of codimension $g - 1$ is more subtle. From the definition of the tautological ring, tautological classes are obtained by taking ψ-classes, and multiplying, gluing, and pushing forward by forgetful morphisms. Now on

$\mathcal{M}_g = \mathcal{M}_{g,0}$, there are no ψ-classes and no boundary strata, so by the definition of the tautological ring, all codimension $g-1$ tautological classes on $\mathcal{M}_{g,0}$ are pushed forward from tautological classes on $\mathcal{M}_{g,1}$, which are necessarily of codimension g. These also vanish by Theorem $\star$ by the same argument as before.

As before, one can say more:

Exercise. Extend this argument to the moduli space of curves with rational tails $\mathcal{M}_{g,n}^{rt}$. (First determine the dimension of the conjectural Poincaré duality ring!)

The Faber-type conjecture for this space was mentioned in Sect. 3.24. I should point out that I expect that Looijenga's proof extends to this case without problem, but I haven't checked.

4.6. *Theorem $\star$ implies the socle part of the Hain–Looijenga–Faber–Pandharipande Conjecture 3.11 on $\overline{\mathcal{M}}_{g,n}$.* Recall the socle part of the Hain–Looijenga–Faber–Pandharipande Conjecture 3.11, that $R_0(\overline{\mathcal{M}}_{g,n}) \cong \mathbb{Q}$.

We show how this is implied by Theorem $\star$. This was first shown in [GrV2], which can be seen as a first step toward the statement and proof of Theorem $\star$.

Our goal is to show that all tautological 0-cycles are commensurate, and that one of them is non-zero. Clearly the latter is true, as the class of a 0-dimensional stratum (a point) is tautological, and is non-zero as it has non-zero degree, so we concentrate on the first statement.

By Theorem $\star$, any dimension 0 tautological class is pushed forward from the locus of curves with at least $(3g-3+n)-g+1 = 2g-2+n$ genus 0 components.

Exercise. Show that the only stable dual graphs with $2g-2+n$ genus 0 components have all vertices genus 0 and trivalent. Show that these are the 0-dimensional strata. (See Fig. 7 for the 0-dimensional strata on $\overline{\mathcal{M}}_{1,2}$.)

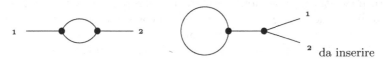
da inserire

Fig. 7. The 0-dimensional strata on $\overline{\mathcal{M}}_{1,2}$ – notice that all vertices are genus 0 and trivalent, and that there are $2g-2+n$ of them

Hence $R_0(\overline{\mathcal{M}}_{g,n})$ is generated by this finite number of points. It remains to show that any two of these points are equivalent in the Chow ring. A geometric way of showing this is by observing that all points in $\overline{\mathcal{M}}_{0,N}$ are equivalent in the Chow ring, and that our 0-dimensional strata are in the image of $\overline{\mathcal{M}}_{0,2g+n}$ under g gluing morphisms. A more combinatorial way of showing this is by showing that each 1-dimensional stratum is isomorphic to $\mathbb{P}^1$, and that any two 0-dimensional strata can be connected by a chain of 1-dimensional strata.

Exercise. Complete one of these arguments.

As in the earlier applications of Theorem $\star$ too: we can verify the perfect pairing conjecture in codimension 1 and probably 2 (although Tom Graber and I haven't delved too deeply into 2). This is combinatorially more serious, but not technically hard.

4.7. *Theorem $\star$ implies the Faber–Pandharipande vanishing/socle conjecture on curves of compact type.* We now show the "vanishing/socle part" of the Faber-type conjecture for curves of compact type (Faber–Pandharipande Conjecture 3.26).

First, suppose that $i > 2g - 3$. We will show that $R^i(\mathcal{M}_g^c) = 0$. By Theorem $\star$, any such tautological class vanishes on the open set where there are at most $i - g + 1 > g - 2$ genus 0 vertices. Then our goal follows from the next exercise.

Exercise. Show that any genus g (0-pointed) stable graph that is a tree has at most $g - 2$ genus 0 vertices. Moreover, if equality holds, then each vertex is either genus 1 of valence 1, or genus 0 of valence 3. (Examples when $g = 6$ are given in Fig. 8.)

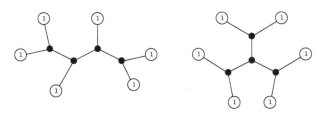

Fig. 8. The two 0-pointed genus 6 stable trees with at least 4 genus 0 vertices

Next, if $i = 2g - 3$, then our codimension $2g - 3$ (hence g) class is pushed forward from strata of the form described in the previous exercise. But each stratum has dimension g, so the tautological class must be a linear combination of fundamental classes of such strata.

Furthermore, any two such strata are equivalent (in cohomology, or even in the Chow ring) by arguments analogous to either of those we used for $\overline{\mathcal{M}}_{g,n}$.

Thus we have shown that $R^{2g-3}(\mathcal{M}_g^c)$ is generated by the fundamental class of a single such stratum. It remains to show that this is non-zero. This argument is short, but requires a little more background than we have presented. (For the experts: it suffices to show that $\lambda_g \neq 0$ on this stratum $\mathcal{M}_\Gamma$. We have a cover $\pi\overline{\mathcal{M}}_{1,1}^g \to \mathcal{M}_\Gamma$ via gluing morphisms, and the pullback of the Hodge bundle splits into the Hodge bundles of each of the g elliptic curves. Thus $\pi^*\lambda_g$ is the product of the λ_1-classes on each factor, so $\deg \pi^*\lambda_g = (\int_{\overline{\mathcal{M}}_{1,1}} \lambda_1)^g = 1/24^g \neq 0$.)

As always, Theorem $\star$ gives extra information. (1) This argument extends to curves of compact type with points. (2) We can now attack part of the

Poincaré duality portion of the conjecture. (3) We get an explicit generator of $R^{2g-3}(\mathcal{M}_g^c)$ (a stratum of a particular form, e.g. Fig. 8).

4.8. *Theorem $\star$ helps determine the tautological ring in low dimension.* In the course of proving $R_0(\overline{\mathcal{M}}_{g,n}) \cong \mathbb{Q}$, we showed that $R_0(\overline{\mathcal{M}}_{g,n})$ was generated by 0-strata. A similar argument shows that $R_i(\overline{\mathcal{M}}_{g,n})$ generated by boundary strata for $i = 1, 2$ (where $R_i(\overline{\mathcal{M}}_{g,n}) := R^{3g-3+n-i}(\overline{\mathcal{M}}_{g,n})$). (We are already aware that this will not extend to $i = 3$, as ψ_1 on $\overline{\mathcal{M}}_{2,1}$ is not a linear combination of fundamental classes of strata.)

In general, Theorem $\star$ implies that in order to understand tautological classes in dimension up to i, you need only understand curves of genus up to $(i+1)/2$, with not too many marked points.

The moral of this is that the "top" (lowest-codimension) part of the tautological ring used to be considered the least mysterious (given the definition of the tautological ring, it is easy to give generators), and the bottom was therefore the most mysterious. Now the situation is the opposite. For example, in codimension 3, we can describe the generators of the tautological ring, but we have no idea what the relations are. However, we know exactly what the tautological ring looks like in *dimension* 3.

4.9. *Exercise.* Use Theorem $\star$ and a similar argument to show that the tautological groups of $\overline{\mathcal{M}}_{0,n}$ and $\overline{\mathcal{M}}_{1,n}$ are generated by boundary strata.

4.10. *Additional consequences.* For many additional consequences of Theorem $\star$, see [GrV3]. For example, we recover *Diaz' theorem* ($\mathcal{M}_g$ contains no complete subvarieties of dim $> g - 2$), as well as generalizations and variations such as: $\overline{\mathcal{M}}_{g,n}^c$ contains no complete subvarieties of dim $> 2g - 3 + n$.

The idea behind the proof of Theorem $\star$ is rather naive. But before we can discuss it, we will have to finally enter the land of Gromov–Witten theory, and define stable relative maps to $\mathbb{P}^1$, which we will interpret as a generalization of the notion of a branched cover.

5 Stable Relative Maps to $\mathbb{P}^1$ and Relative Virtual Localization

We now discuss the theory of *stable relative maps*, and "virtual" localization on their moduli space (*relative virtual localization*). We will follow J. Li's algebro-geometric definition of stable relative maps [Li1], and his description of their obstruction theory [Li2], but we point out earlier definitions of stable relative maps in the differentiable category due to A.-M. Li and Y. Ruan [LR], and Ionel and Parker [IP1, IP2], and Gathmann's work in the algebraic category in genus 0 [Ga]. We need the algebraic category for several reasons, most importantly because we will want to apply virtual localization.

Stable relative maps are variations of the notion of stable maps, and the reader may wish to become comfortable with that notion first. (Stable maps are discussed in Abramovich's article in this volume, for example.)

We are interested in the particular case of stable relative maps to $\mathbb{P}^1$, relative to at most two points, so we will define stable relative maps only in this case. For concreteness, we define stable maps to $X = \mathbb{P}^1$ relative to one point ∞; the case of zero or two points is the obvious variation on this theme. Such a stable relative map to $(\mathbb{P}^1, \infty)$ is defined as follows. We are given the data of a degree d of the map, a genus g of the source curve, a number m of marked points, and a partition $d = \alpha_1 + \cdots + \alpha_n$, which we write $\alpha \vdash d$.

Then a relative map is the following data:

- A morphism f_1 from a nodal $(m + n)$-pointed genus g curve $(C, p_1, \ldots, p_m, q_1, \ldots, q_n)$ (where as usual the p_i and q_j are distinct non-singular points) to a chain of $\mathbb{P}^1$'s, $T = T_0 \cup T_1 \cup \cdots \cup T_t$ (where T_i and T_{i+1} meet), with a point $\infty \in T_t - T_{t-1}$. Unfortunately, there are two points named ∞. We will call the one on X, ∞_X, and the one on T, ∞_T, whenever there is any ambiguity.
- A projection $f_2 : T \to X$ contracting T_i to ∞_X (for $i > 0$) and giving an isomorphism from $(T_0, T_0 \cap T_1)$ (resp. (T_0, ∞)) to X if $t > 0$ (resp. if $t = 0$). Denote $f_2 \circ f_1$ by f.
- We have an equality of divisors on C: $f_1^* \infty_T = \sum \alpha_i q_i$. In particular, $f_1^{-1} \infty_T$ consists of nonsingular (marked) points of C.
- The preimage of each node n of T is a union of nodes of C. At any such node n' of C, the two branches map to the two branches of n, and their orders of branching are the same. (This is called the *predeformability* or *kissing* condition.)

If follows that the degree of f_1 is d on each T_i. An *isomorphism* of two such maps is a commuting diagram

$$
\begin{array}{ccc}
(C, p_1, \ldots, p_m, q_1, \ldots, q_n) & \xrightarrow{\sim} & (C', p_1', \ldots, p_m', q_1', \ldots, q_n') \\
\downarrow f_1 & & \downarrow f_1 \\
(T, \infty_T) & \xrightarrow{\sim} & (T, \infty_T) \\
\downarrow f_2 & & \downarrow f_2 \\
(X, \infty_X) & \xrightarrow{=} & (X, \infty_X)
\end{array}
$$

where all horizontal morphisms are isomorphisms, the bottom (although not necessarily the middle!) is an equality, and the top horizontal isomorphism sends p_i to p_i' and q_j to q_j'. Note that the middle isomorphism must preserve the isomorphism of T_0 with X, and is hence the identity on T_0, but for $i > 0$, the isomorphism may not be the identity on T_i.

This data of a relative map is often just denoted f, with the remaining information left implicit.

We say that f is *stable* if it has finite automorphism group. This corresponds to the following criteria:

- Any f_1-contracted geometric genus 0 component has at least 3 "special points" (node branches or marked points).
- Any f_1-contracted geometric genus 1 component has at least 1 "special point".
- If $0 < i < t$ (resp. $0 < i = t$), then not every component mapping to T_i is of the form $[x; y] \to [x^q; y^q]$, where the coordinates on the target are given by $[0; 1] = T_i \cap T_{i-1}$ and $[1; 0] = T_i \cap T_{i+1}$ (resp. $[1; 0] = \infty$).

(The first two conditions are the same as for stable maps. The third condition is new.) A picture of a stable relative map is given in Fig. 9.

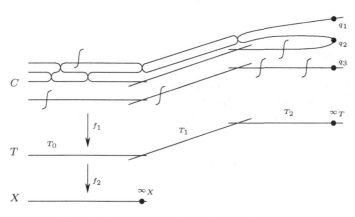

Fig. 9. An example of a stable relative map

Thus we have some behavior familiar from the theory of stable maps: we can have contracted components, so long as they are "stable", and don't map to any nodes of T, or to ∞_T. We also have some new behavior: the target X can "sprout" a chain of $\mathbb{P}^1$'s at ∞_X. Also, the action of $\mathbb{C}^*$ on the map via the action on a component T_i $(i > 0)$ that preserves the two "special points" of T_i (the intersections with T_{i-1} and T_{i+1} if $i < t$, and the intersection with T_{i-1} and ∞ if $i = t$) is considered to preserve the stable map. For example, Fig. 10 shows two isomorphic stable maps.

There is a compact moduli space (Deligne–Mumford stack) for stable relative maps to $\mathbb{P}^1$, denoted $\overline{\mathcal{M}}_{g,m,\alpha}(\mathbb{P}^1, d)$. (In order to be more precise, I should tell you the definition of a family of stable relative maps parametrized by an arbitrary base, but I will not do so.) In what follows, $m = 0$, and that subscript will be omitted. (More generally, stable relative maps may be defined with $\mathbb{P}^1$ replaced by any smooth complex projective variety, and the point ∞ replaced by any smooth divisor D on X. The special case $D = \emptyset$ yields Kontsevich's original space of stable maps.)

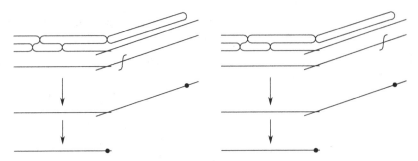

Fig. 10. Two isomorphic stable relative maps

Unfortunately, the space $\overline{\mathcal{M}}_{g,\alpha}(\mathbb{P}^1, d)$ is in general terribly singular, and not even equidimensional.

Exercise. Give an example of such a moduli space with two components of different dimensions. (Hint: use contracted components judiciously.)

However, it has a component which we already understand well, which corresponds to maps from a smooth curve, which is a branched cover of $\mathbb{P}^1$. Such curves form a moduli space $\mathcal{M}_{g,\alpha}(\mathbb{P}^1, d)$ of dimension corresponding to the "expected number of branch points distinct from ∞", which we may calculate by the Riemann–Hurwitz formula (2) to be

$$(10) \qquad\qquad r = 2g - 2 + n + d.$$

We have seen this formula before, (6).

Exercise. Verify (10).

These notions can be readily generalized, for example to stable relative maps to $\mathbb{P}^1$ relative to two points (whose moduli space is denoted $\overline{\mathcal{M}}_{g,\alpha,\beta}(\mathbb{P}^1, d)$), or to no points (otherwise known as the stable maps to $\mathbb{P}^1$; this moduli space is denoted $\overline{\mathcal{M}}_g(\mathbb{P}^1, d)$).

Exercise. Calculate $\dim \mathcal{M}_{g,\alpha,\beta}(\mathbb{P}^1, d)$ (where α has m parts and β has n parts) and $\dim \mathcal{M}_g(\mathbb{P}^1, d)$.

5.1. *Stable relative maps with possibly-disconnected source curve.* Recall that by our (non-standard) definition, nodal curves are connected. It will be convenient, especially when discussing the degeneration formula, to consider curves without this hypothesis. Just as our discussion of (connected) stable curves generalized without change to (possibly-disconnected) stable curves (see Sect. 3.30), our discussion of (relatively) stable maps from connected curves generalizes without change to "(relatively) stable maps from possibly-disconnected curves". Let $\overline{\mathcal{M}}_{g,\alpha}(\mathbb{P}^1, d)^{\bullet}$ be the space of stable relative maps from possibly-disconnected curves (to $\mathbb{P}^1$, of degree d, etc.). Warning: this is *not* in general the quotient of a product of $\overline{\mathcal{M}}_{g',\alpha'}(\mathbb{P}^1, d')$'s by a finite group.

5.2. The Virtual Fundamental Class

There is a natural homology (or Chow) class on $\overline{\mathcal{M}}_{g,\alpha}(\mathbb{P}^1, d)$ of dimension $r = \dim \mathcal{M}_{g,\alpha}(\mathbb{P}^1, d)$ (cf. (10)), called the *virtual fundamental class* $[\overline{\mathcal{M}}_{g,\alpha}(\mathbb{P}^1, d)]^{\mathrm{virt}} \in A_r(\overline{\mathcal{M}}_{g,\alpha}(\mathbb{P}^1, d)])$, which is obtained from the deformation-obstruction theory of stable relative maps, and has many wonderful properties. The virtual fundamental class agrees with the actual fundamental class on the open subset $\mathcal{M}_{g,\alpha}(\mathbb{P}^1, d)$. The most difficult part of dealing with the moduli space of stable relative maps is working with the virtual fundamental class.

Aside: relative Gromov–Witten invariants. In analogy with usual Gromov–Witten invariants, one can define *relative Gromov–Witten invariants* by intersecting natural cohomology classes on the moduli space with the virtual fundamental class. More precisely, one multiplies (via the cup/cap product) the cohomology classes with the virtual fundamental class, and takes the degree of the resulting zero-cycle. One can define ψ-classes and λ-classes in the same way as before, and include these in the product. When including ψ-classes, the numbers are often called *descendant relative invariants*; when including λ-classes, the numbers are sometimes called *Hodge integrals*. For example, one can show that Hurwitz numbers are descendant relative invariants of $\mathbb{P}^1$. However, this point of view turns out to be less helpful, and we will not use the language of relative Gromov–Witten invariants again.

The virtual fundamental class behaves well under two procedures: degeneration and localization; we now discuss these.

5.3. The Degeneration Formula for the Virtual Fundamental Class, Following [Li2]

We describe the degeneration formula in the case of stable maps to $\mathbb{P}^1$ relative to one point, and leave the cases of stable maps to $\mathbb{P}^1$ relative to zero or two points as straightforward variations for the reader. In this discussion, we will deal with possibly-disconnected curves to simplify the exposition.

Consider the maps to $\mathbb{P}^1$ relative to one point ∞, and imagine deforming the target so that it breaks into two $\mathbb{P}^1$'s, meeting at a node (with ∞ on one of the components). It turns out that the virtual fundamental class behaves well under this degeneration. The limit can be expressed in terms of virtual fundamental classes of spaces of stable relative maps to each component, relative to ∞ (for the component containing ∞), and relative to the node-branch (for both components).

Before we make this precise, we give some intuition. Suppose we have a branched cover $C \to \mathbb{P}^1$, and we deform the target into a union of two $\mathbb{P}^1$'s, while keeping the branch points away from the node; call the limit map $C' \to \mathbb{P}^1 \cup \mathbb{P}^1$. Clearly in the limit, away from the node, the cover looks just the same as it did before (with the same branching). At the node, it turns out that the branched covers of the two components must satisfy the kissing/predeformability condition. Say that the branching above the node corresponds to the partition $\gamma_1 + \cdots + \gamma_m$. By our discussion about Hurwitz numbers, as we have specified the branch points, there will be a finite number of such branched covers – we count branched covers of each component of $\mathbb{P}^1 \cup \mathbb{P}^1$, with branching corresponding to the partition γ above the node-branch; then we choose how to match the preimages of the node-branch on the two components (there are $\#\text{Aut}\gamma$ such choices). It turns out that $\gamma_1 \cdots \gamma_m$ covers of the original sort will degenerate to each branched cover of the nodal curve of this sort. (Notice that if we were interested in connected curves C, then the inverse image of each component of $\mathbb{P}^1$ would not necessarily be connected, and we would have to take some care in gluing these curves together to get a connected union. This is the reason for considering possibly-disconnected components.)

Motivated by the previous paragraph, we give the degeneration formula. Consider the degeneration of the target $(\mathbb{P}^1, \infty) \rightsquigarrow \mathbb{P}^1 \cup (\mathbb{P}^1, \infty)$. Let (X, ∞) be the general target, and let (X', ∞) be the degenerated target. Let $(X_1, a_1) \cong (\mathbb{P}^1, \infty)$ denote the first component of X', where a_1 refers to the node-branch, and let $(X_2, a_2, \infty) \cong (\mathbb{P}^1, 0, \infty)$ denote the second component of X', where a_2 corresponds to the node branch. Then for each partition $\gamma_1 + \cdots + \gamma_m = d$, there is a natural map

$$(11) \qquad \overline{\mathcal{M}}_{g_1,\gamma}(\mathbb{P}^1, d)^\bullet \times \overline{\mathcal{M}}_{g_2,\gamma,\alpha}(\mathbb{P}^1, d)^\bullet \to \overline{\mathcal{M}}_{g_1+g_2-m+1,\alpha}(X', d)^\bullet$$

obtained by gluing the points above a_1 to the corresponding points above a_2. The image of this map can be suitably interpreted as stable maps to X', satisfying the kissing condition, which can appear as the limit of maps to X. (We are obscuring a delicate issue here – we have not defined stable maps to a singular target such as X.) Then Li's degeneration formula states that the image of the product of the virtual fundamental classes in (11) is the limit of the virtual fundamental class of $\overline{\mathcal{M}}_{g,\alpha}(\mathbb{P}^1, d)^\bullet$, multiplied by $\gamma_1 \cdots \gamma_m$.

The main idea behind Li's proof is remarkably elegant, but as with any argument involving the virtual fundamental class, the details are quite technical.

If we are interested in connected curves, then there is a corresponding statement (that requires no additional proof): we look at the component of the moduli space on the right side of (11) corresponding to maps from connected source curves, and we look at just those components of the moduli spaces on the left side which glue together to give connected curves.

5.4. Relative Virtual Localization [GrV3]

The second fundamental method of manipulating virtual fundamental classes is by means of localization. Before discussing localization in our Gromov–Witten-theoretic context, we first quickly review localization in its original setting.

(A friendly introduction to equivariant cohomology is given in [HKKPTVVZ, Chap. 4], and to localization on the space of ordinary stable maps in [HKKPTVVZ, Chap. 27].)

Suppose Y is a complex projective manifold with an action by a torus $\mathbb{C}^*$. Then the fixed point locus of the torus is the union of smooth submanifolds, possibly of various dimensions. Let the components of the fixed locus be Y_1, Y_2, ... The torus acts on the normal bundle N_i to Y_i. Then the Atiyah-Bott localization formula states that

$$(12) \qquad [Y] = \sum_{\text{fixed}} [Y_i]/c_{\text{top}}(N_i) = \sum_{\text{fixed}} [Y_i]/e(N_i),$$

in the *equivariant homology* of Y (with appropriate terms inverted), where c_{top} (or the Euler class e) of a vector bundle denotes the top Chern class. This is a wonderfully powerful fact, and to appreciate it, you must do examples yourself. The original paper of Atiyah and Bott [AB] is beautifully written and remains a canonical source.

You can cap (12) with various cohomology classes to get 0-dimensional classes, and get an equality of numbers. But you can cap (12) with classes to get higher-dimensional classes, and get equality in cohomology (or the Chow ring). One lesson I want to emphasize is that this is a powerful thing to do. For example, in a virtual setting, in Gromov–Witten theory, localization is traditionally used to get equalities of numbers. We will also use equalities of numbers to prove the ELSV formula (8). However, using more generally equalities of classes will give us Theorem $\star$, and part of Faber's conjecture.

Localization was introduced to Gromov–Witten theory by Kontsevich in his ground-breaking paper [Ko2], in which he works on the space of genus zero maps to projective space, where the virtual fundamental class is the usual fundamental class (and hence there are no "virtual" technicalities). In the foundational paper [GrP], Graber and Pandharipande showed that the localization formula (12) works "virtually" on the moduli space of stable maps, where fundamental classes are replaced by virtual fundamental classes, and normal bundles are replaced by "virtual normal bundles". They defined the virtual fundamental class of a fixed locus, and the virtual normal bundle, and developed the machinery to deal with such questions.

There is one pedantic point that must be made here. The localization formula should reasonably be expected to work in great generality. However, we currently know it only subject to certain technical hypotheses. (1) The proof only works in the algebraic category. (2) In order to apply this machinery, the moduli space must admit a $\mathbb{C}^*$-equivariant locally closed immersion into an orbifold. (3) The virtual fundamental class of this fixed locus needs to be shown to arise from the $\mathbb{C}^*$-fixed part of the obstruction theory of the moduli space. It would be very interesting, and potentially important, to remove hypotheses (1) and (2).

The theory of virtual localization can be applied to our relative setting [GrV3]. (See [LLZ] for more discussion.) We now describe it in the case of interest to us, of maps to $\mathbb{P}^1$. Again, in order to understand this properly, you should work out examples yourself.

Fix a torus action on $\mathbb{P}^1$

$$\sigma \circ [x; y] = [\sigma x; y],$$

so the torus acts with weight 1 on the tangent space at 0 and -1 on the tangent space at ∞. (The *weight* is the one-dimensional representation, or equivalently, the character.) This torus action induces an obvious torus action on $\overline{\mathcal{M}}_{g,\alpha}(\mathbb{P}^1, d)^\bullet$ (and $\overline{\mathcal{M}}_{g,\alpha}(\mathbb{P}^1, d)$).

We first determine the torus-fixed points of this action. Suppose $C \to T \to X$ is such a fixed map. A picture of two fixed maps showing "typical" behavior is given in Fig. 11. The first has "nothing happening above ∞_X", and the second has some "sprouting" of T_i's.

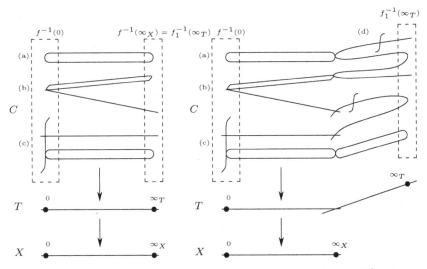

Fig. 11. Two examples of torus-fixed stable relative maps to $(\mathbb{P}^1, \infty)$

The map $C \to X$ must necessarily be a covering space away from the points 0 and ∞ of $X = \mathbb{P}^1$.

Exercise. Using the Riemann–Hurwitz formula, show that a surjective map $C' \to \mathbb{P}^1$ from an irreducible curve, unbranched away from 0 and ∞, must be of the form $\mathbb{P}^1 \to \mathbb{P}^1$, $[x; y] \mapsto [x^a; y^a]$ for some a.

5.5. Hence the components dominating X must be a union of "trivial covers" of this sort.

We now focus our attention on the preimage of 0. Any sort of (stable) behavior above 0 is allowed. For example, the curve could be smooth and branched there (Fig. 11a); or two of the trivial covers could meet in a node (Fig. 11b); or there could be a contracted component of C, intersecting various trivial components at nodes (Fig. 11c). (Because the "relative" part of the picture is at ∞, this discussion is the same as the discussion for ordinary stable maps, as discussed in [GrP].)

Finally, we consider the preimage of ∞_X. Possibly "nothing happens over ∞", i.e. the target has not sprouted a tree ($l = 0$ in our definition of stable relative maps at the start of Sect. 5), and the preimage of ∞ consists just of n smooth points; this is the first example in Fig. 11. Otherwise, there is some "sprouting" of the target, and something "nontrivial" happens above each sprouted component T_i $(i > 0)$, as in Fig. 11d.

5.6. At this point, you should draw some pictures, and convince yourself of the following important fact: the connected components of the fixed locus correspond to certain discrete data. In particular, each connected component can be interpreted as a product of three sorts of moduli spaces:

(A) moduli spaces of pointed curves (corresponding to Fig. 11c)
(B) (for those fixed loci where "something happens above ∞_X, i.e. Fig. 11d), a moduli space of maps parametrizing the behaviour there. This moduli space is a variant of the space of stable relative maps, where there is no "rigidifying" map to X. We denote such a moduli space by $\overline{\mathcal{M}}_{g,\alpha,\beta}(\mathbb{P}^1, d)_\sim$; its theory (of deformations and obstructions and virtual fundamental classes) is essentially the same as that for $\overline{\mathcal{M}}_{g,\alpha,\beta}(\mathbb{P}^1, d)$. The virtual dimension of $\overline{\mathcal{M}}_{g,\alpha,\beta}(\mathbb{P}^1, d)_\sim$ is one less than that of $\overline{\mathcal{M}}_{g,\alpha,\beta}(\mathbb{P}^1, d)$.
(C) If $\alpha_1 + \cdots + \alpha_n = d$ is the partition corresponding to the "trivial covers" of T_0, these stable relative maps have automorphisms $\mathbb{Z}_{\alpha_1} \times \cdots \times \mathbb{Z}_{\alpha_n}$ corresponding to automorphisms of these trivial covers (i.e. if one trivial cover is of the form $[x; y] \mapsto [x^{\alpha_1}; y^{\alpha_1}]$, and ζ_{α_1} is a α_1th root of unity, then $[x; y] \mapsto [\zeta_{\alpha_1} x; y]$ induces an automorphism of the map). In the language of stacks, we can include a factor of $B\mathbb{Z}_{\alpha_1} \times \cdots \times B\mathbb{Z}_{\alpha_n}$; but the reader may prefer to simply divide the virtual fundamental class by $\prod \alpha_i$ instead.

Each of these spaces has a natural virtual fundamental class: the first sort has its usual fundamental class, and the second has its intrinsic virtual fundamental class.

The *relative virtual localization* formula states that

$$[\overline{\mathcal{M}}_{g,\alpha}(\mathbb{P}^1, d)]^{\text{virt}} = \sum_{\text{fixed}} [Y_i]^{\text{virt}}/e(N_i^{\text{virt}}),$$

in the equivariant homology of $\overline{\mathcal{M}}_{g,\alpha}(\mathbb{P}^1, d)$ (cf. (12)), with suitable terms inverted, where the virtual fundamental classes of the fixed loci are as just described, and the "virtual normal bundle" will be defined now.

Fix attention now to a fixed component Y_i. The virtual normal bundle is a class in equivariant K-theory. The term $1/e(N_i^{\text{virt}})$ can be interpreted as the product of several factors, each "associated" to a part of the picture in Fig. 11. We now describe these contributions. The reader is advised to not worry too much about the precise formulas; the most important thing is to get a sense of the shape of the formula upon a first exposure to these ideas. Let t be the generator of the equivariant cohomology of a point (i.e. $H_T^*(\text{pt}) = \mathbb{Z}[t]$):

1. For each irreducible component dominating T_0 (i.e. each trivial cover) of degree α_i, we have a contribution of $\frac{\alpha_i^{\alpha_i}}{\alpha_i! t^{\alpha_i}}$.

2. For each contracted curve above 0 (Fig. 11c) of genus g', we have a contribution of $(t^{g'} - \lambda_1 t^{g'-1} + \cdots + (-1)^{g'} \lambda_{g'})/t$. (This contribution is on the factor $\overline{\mathcal{M}}_{g',n}$ corresponding to the contracted curve.)

3. For each point where a trivial component of degree α_i meets a contracted curve above 0 at a point j, we have a contribution of $t/(t/\alpha_i - \psi_j)$. here, ψ_j is a class on the moduli space $\overline{\mathcal{M}}_{g',n}$ corresponding to the contracted component.

4. For each node above 0 (Fig. 11b) joining trivial covers of degrees α_i and α_j, we have a contribution of $1/(t/\alpha_1 + t/\alpha_2)$.

5. For each smooth point above 0 (Fig. 11a) on a trivial cover of degree α_i, we have a contribution of t/α_i.

At this point, if you squint and ignore the t's, you can almost see the ELSV formula (8) taking shape.

6. If there is a component over ∞_X, then we have a contribution of $1/(-t - \psi)$, where ψ is the first Chern class of the line bundle corresponding to the cotangent space of T_1 at the point where it meets T_0.

These six contributions look (and are!) complicated. But this formula can be judiciously used to give some powerful results, surprisingly cheaply. We now describe some of these.

6 Applications of Relative Virtual Localization

6.1. Example 1: Proof of the ELSV Formula (8)

As a first example, we prove the ELSV formula (8). (This formula follows [GrV1], using the simplification in the last section of [GrV1] provided by the existence of Jun Li's description of the moduli space of stable relative maps.) The ELSV formula counts branched covers with specified branching over ∞ corresponding to $\alpha \vdash d$, and other fixed simple branched points. Hence we will consider $\overline{\mathcal{M}}_{g,\alpha}(\mathbb{P}^1, d)$.

We next need to impose other fixed branch points. There is a natural Gromov–Witten-theoretic approach involving using descendant invariants, but this turns out to be the wrong thing to do. Instead, we use a beautiful construction of Fantechi and Pandharipande [FanP]. Given any map from a nodal curve to $\mathbb{P}^1$, we can define a *branch divisor* on the target. When the source curve is smooth, the definition is natural (and old): above a point p corresponding to a partition $\beta \vdash d$, the branch divisor contains p with multiplicity $\sum(\beta_i - 1)$. It is not hard to figure out how extend this to the case where the source curve is not smooth above p.

Exercise. Figure out what this extension should be. (Do this so that the Riemann–Hurwitz formula remains true.)

Thus we have a map of sets $\overline{\mathcal{M}}_g(\mathbb{P}^1, d) \to \operatorname{Sym}^{2d+2g-2}\mathbb{P}^1$. In the case of stable relative maps, we have a map of sets $\overline{\mathcal{M}}_{g,\alpha}(\mathbb{P}^1, d) \to \operatorname{Sym}^{2d+2g-2}\mathbb{P}^1$. As each such stable relative map will have branching of at least $\sum(\alpha_i - 1)$ above ∞, we can subtract this fixed branch divisor to get a map of sets

$$(13) \qquad\qquad br : \overline{\mathcal{M}}_{g,\alpha}(\mathbb{P}^1, d) \to \mathrm{Sym}^r \mathbb{P}^1$$

where $r = 2d + 2g - 2 - \sum(\alpha_i - 1) = 2g - 2 + d + n$ (cf. (10)).

The important technical result proved by Fantechi and Pandharipande is the following.

6.2. *Theorem (Fantechi–Pandharipande [FanP]). There is a natural map of stacks br as in* (13).

We call such a map a *(Fantechi–Pandharipande) branch morphism*. This morphism respects the torus action.

One can now readily verify several facts. If the branch divisor does not contain $p \neq \infty$ in $\mathbb{P}^1$, then the corresponding map $C \to \mathbb{P}^1$ is unbranched (i.e. a covering space, or étale) above p. If the branch divisor contains $p \neq \infty$ with multiplicity 1, then the corresponding map is simply branched above p. (Recall that this means that the preimage of p consists of smooth points, and the branching corresponds to the partition $2 + 1 + \cdots + 1$.) If the branch divisor does not contain ∞, i.e. there is no additional branching above ∞_X beyond that required by the definition of stable relative map, then the preimage of ∞_X consists precisely of the n smooth points q_i. In other words, there is no "sprouting" of T_i, i.e. $T \cong \mathbb{P}^1$. Hence if $p_1 + \cdots + p_r$ is a general point of $\mathrm{Sym}^r \mathbb{P}^1$, then $br^{-1}(p_1 + \cdots + p_r) \subset \overline{\mathcal{M}}_{g,\alpha}(\mathbb{P}^1, d)$ is a finite set of cardinality equal to the Hurwitz number H_α^g. This is true despite the fact that $\overline{\mathcal{M}}_{g,\alpha}(\mathbb{P}^1, d)$ is horribly non-equidimensional – the preimage of a general point of $\mathrm{Sym}^r \mathbb{P}^1$ will be contained in $\mathcal{M}_{g,\alpha}(\mathbb{P}^1, d)$, and will not meet any other nasty components!

By turning this set-theoretic argument into something more stack-theoretic and precise, we have that

$$(14) \qquad\qquad H_\alpha^g = \deg br^{-1}(pt) \cap [\overline{\mathcal{M}}_{g,\alpha}(\mathbb{P}^1, d)]^{\mathrm{virt}}.$$

(For distracting unimportant reasons, the previous paragraph's discussion is slightly incorrect in the case where $H_\alpha^g = 1/2$, but (14) is true.)

We can now calculate the right side of (14) using localization. In order to do this, we need to interpret it equivariantly, which involves choosing an equivariant lift of br^{-1} of a point in $\mathrm{Sym}^r \mathbb{P}^1 \cong \mathbb{P}^r$. We do this by choosing our point in $\mathrm{Sym}^r \mathbb{P}^1$ to be the point 0 with multiplicity r. Thus all the branching (aside from that forced to be at ∞) must be at 0. The normal bundle to this point of $\mathbb{P}^r$ is $r! t^r$. Thus when we apply localization, a miracle happens. The only fixed loci we consider are those where there is no extra branching over ∞ (see the first picture in Fig. 11). However, the source curve is smooth, so there is in fact only one connected component of the fixed locus to consider, which is shown in Fig. 12. The moduli space in this case is $\overline{\mathcal{M}}_{g,n}$, which we take with multiplicity $1/\prod \alpha_i$ (cf. Sect. 5.6(C)). Hence the Hurwitz number is the intersection on this moduli space of the contributions to the virtual normal bundle outlined above.

Exercise. Verify that the contributions from *1*, *2*, and *3* above, on the moduli space $\overline{\mathcal{M}}_{g,n}$, give the ELSV formula (8).

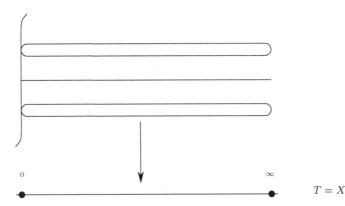

Fig. 12. The only fixed locus contributing to our calculation of the Hurwitz number

6.3. Example 2: Proof of Theorem ⋆ (Theorem 4.1)

In Example 1 (Sect. 6.1), we found an equality of numbers. Here we will use relative virtual localization to get equality of cohomology or Chow classes.

Fix g and n. We are interested in dimension j (tautological) classes on $\overline{\mathcal{M}}_{g,n}$. In particular, we wish to show that any such *tautological* class can be deformed into one supported on the locus corresponding to curves with at least $2g-2+n-j$ genus 0 components. (This is just a restatement of Theorem ⋆.) Call such a dimension j class *good*. Using the definition of the tautological ring in terms of ψ-classes, it suffices to show that monomials in ψ-classes of dimension j (hence degree=codimension $3g-3+n-j$) are good.

Here is one natural way of getting dimension j classes. Take any partition $\alpha_1 + \cdots + \alpha_n = d$. Let $r = 2g-2+n+d$ be the virtual dimension of $\overline{\mathcal{M}}_{g,\alpha}(\mathbb{P}^1, d)$ (i.e. the dimension of the virtual fundamental class, and the dimension of $\mathcal{M}_{g,\alpha}(\mathbb{P}^1, d)$), and suppose $r > j$. Define the *Hurwitz class* $\mathbb{H}_j^{g,\alpha}$ by

$$\mathbb{H}_j^{g,\alpha} := \pi_* \left((\cap_{i=1}^{r-j} br^{-1}(p_i)) \cap [\overline{\mathcal{M}}_{g,\alpha}(\mathbb{P}^1, d)]^{\text{virt}} \right) \in A_j(\overline{\mathcal{M}}_{g,n})$$

where π is the moduli map $\overline{\mathcal{M}}_{g,\alpha}(\mathbb{P}^1, d) \to \overline{\mathcal{M}}_{g,n}$ (and the n points are the preimages of ∞), and $p_1, \ldots, p_{r-j}$ are generally chosen points on $\mathbb{P}^1$. We think of this Hurwitz class informally as follows: consider branched covers with specified branching over ∞. Such covers (and their generalization, stable relative maps) form a space of (virtual) dimension r. Fix all but j branch points, hence giving a class of dimension j. Push this class to the moduli space $\overline{\mathcal{M}}_{g,n}$.

We get at this in two ways, by deformation and by localization.

1. Deformation. (We will implicitly use Li's degeneration formula here.) Deform the target $\mathbb{P}^1$ into a chain of $r-j$ $\mathbb{P}^1$'s, each with one of the fixed branch points p_i. Then you can (and should) check that the stabilized source curve has lots of rational components, essentially as many as stated in Theorem ⋆. (For example, imagine that $r \gg 0$. Then the j "roving" branch points can lie on only a small number of the $r - j$ components of the degenerated target. Suppose $\mathbb{P}^1$ is any other component of the target, where 0 and ∞ correspond to where it meets the previous and next component in the chain. Then the cover restricted to this $\mathbb{P}^1$ can have arbitrary branching over 0 and ∞, and only one other branch point: simple branching above the p_i lying on it. This forces the cover to be a number of trivial covers, plus one other cover $C \to \mathbb{P}^1$, where C is simply branched at p_i, and has one point above 0 and two points above ∞, or vice versa, forcing C to be genus 0, with three node-branches. This analysis will leave you slightly short. The remaining rational curves are found through a clever idea of Ionel, who seemingly conjures a rational curve out of nowhere.) Thus any dimension j Hurwitz class is *good*, i.e. satisfies the conclusion of Theorem $\star$.

2. Localization. We next use localization to express tautological classes in terms of Hurwitz classes. In the same way as for the ELSV formula, we choose an equivariant lifting of $\cap_{i=1}^{r-j} br^{-1}(p_i)$, corresponding to requiring all the p_i to go to 0. (Unlike the ELSV case, there are still j branch points that could go to either 0 or ∞.)

We now consider what fixed components can arise.

We have one "main" component that is similar to the ELSV case, where all the j "roving" branch points go to 0 (Fig. 12). Any other component will be nontrivial over ∞. One can readily inductively show that these other components are *good*, i.e. satisfy the conclusion of Theorem $\star$. (The argument is by looking at the contribution from such a fixed locus. The part contained in $f^{-1}(\infty_X)$ is essentially a Hurwitz class, which we have shown is *good*. The part contained in $f^{-1}(0)$ corresponds to tautological classes on moduli spaces of curves with smaller $2g - 2 + n$, which can be inductively assumed to be *good*.)

Thus we have shown that the contribution of the "main" component is *good*. But this contribution is straightforward to contribute: it is (up to a multiple) the dimension j component of

$$\frac{1 - \lambda_1 + \cdots + (-1)^g \lambda_g}{(1 - \alpha_1 \psi_1) \cdots (1 - \alpha_n \psi_n)}$$

(compare this to the ELSV formula (8)). By expanding this out, we find a polynomial in the α_i of degree $3g - 3 + n - j$ (cf. (9) for a similar argument earlier). We then apply the same trick as when we computed top intersections of ψ-classes using Hurwitz numbers in Sect. 3.20: we can recover the coefficients in this polynomial by "plugging in enough values". In other words, $\psi_1^{a_1} \cdots \psi_n^{a_1}$ may be obtained (modulo *good* classes) as a linear combination of Hurwitz classes. As Hurwitz classes are themselves *good*, we have shown that the monomial $\psi_1^{a_1} \cdots \psi_n^{a_1}$ is also *good*, completing the argument.

7 Towards Faber's Intersection Number Conjecture 3.23 via Relative Virtual Localization

We can use the methods of the proof of Theorem $\star$ to combinatorially describe the top intersections in the tautological ring. Using this, one can prove the "vanishing" or "socle" portion of the Faber-type conjecture for curves with rational tails (and hence for $\mathcal{M}_g$), and prove Faber's intersection number conjecture for up to three points. Details will be given in [GJV3]; here we will just discuss the geometry involved.

The idea is as follows. We are interested in the Chow ring of $\mathcal{M}_{g,n}^{rt}$, so we will work on compact moduli spaces, but discard any classes that vanish on the locus of curves with rational tails. We make a series of short geometric remarks.

First, note that $R_{2g-1}(\mathcal{M}_{g,n}^{rt}) \to R_{2g-1}(\mathcal{M}_{g,1})$ is an isomorphism, and $R_{2g-1}(\mathcal{M}_{g,1}) \to R_{2g-1}(\mathcal{M}_g)$ is a surjection. The latter is immediate from our definition. The argument for the former is for example [GrV3, Prop. 5.8], and can be taken as an exercise for the reader using Theorem $\star$. Faber showed [Fab1, Thm. 2] that $R_{2g-1}(\mathcal{M}_g)$ is non-trivial, so if we can show that $R_{2g-1}(\mathcal{M}_{g,1})$ is generated by a single element, then we will have proved that $R_{2g-1}(\mathcal{M}_{g,n}^{rt}) \cong \mathbb{Q}$ for all $n \geq 0$.

7.1. An extension of that argument using Theorem $\star$ shows that if we have a Hurwitz class of dimension less than $2g - 1$ (i.e. with fewer than $2g - 1$ "moving branch points"), then the class is 0 in $A_*(\mathcal{M}_{g,n}^{rt})$.

In order to get a hold of $R_{2g-1}(\mathcal{M}_{g,n}^{rt})$, we will again use branched covers. Before getting into the Gromov–Witten theory, we make a series of remarks, that may be verified by the reader, using only the Riemann–Hurwitz formula (2).

7.2. First, suppose we have a map $C \to \mathbb{P}^1$ from a nodal (possibly disconnected) curve, unbranched away from 0 and ∞. Then it is a union of trivial covers (in the sense of Sect. 5.5).

7.3. Second, suppose we have a map from a nodal curve C to $\mathbb{P}^1$, with no branching away from 0 and ∞ except for simple branching over 1, and nonsingular over 0 and ∞. Then it is a union of trivial covers, plus one more component, that is genus 0, completely branched over one of $\{0, \infty\}$, and with two preimages over the other. More generally, suppose we have a map from some curve C to a chain of $\mathbb{P}^1$'s, satisfying the kissing condition, unbranched except for two smooth points 0 and ∞ on the ends of the chain, and simple branching over another point 1. Then the map is the union of a number of trivial covers glued together, plus one other cover $\mathbb{P}^1 \to \mathbb{P}^1$ of the component containing 1, of the sort described in the previous sentence.

7.4. Third, if we have a map from a nodal curve C to $\mathbb{P}^1$, with total branching away from 0 and ∞ of degree less than $2g$, and nonsingular over 0 and ∞, then C has no component of geometric genus g. In the same situation, if the total branching away from 0 and ∞ is exactly $2g$, and C has a component of geometric genus g, then the cover is a disjoint union of trivial covers, and one connected curve C' of arithmetic genus g, where the map $C' \to \mathbb{P}^1$ is contracted to 1 or completely branched over 0 and ∞.

More generally, if we have a map from a curve C to a chain of $\mathbb{P}^1$'s satisfying the kissing condition, with 0 and ∞ points on either ends of the chain, with total branching away less than $2g$ away from 0, ∞, and the nodes, then C has no component of geometric genus g. In the same situation, if the total branching away from 0, ∞, and the nodes is precisely $2g$, then the map is the union of a number of trivial covers glued together, plus one other cover of the sort described in the previous paragraph.

7.5. The following fact is trickier. Let $\mathbb{G}_{g,d}$ be the image in $A_{2g-1}(\mathcal{M}_{g,1})$ of $br^{-1}(1) \cap [\overline{\mathcal{M}}_{g,(d),(d)}(\mathbb{P}^1, d)]^{\mathrm{virt}}$ (where the point $p \in C$ corresponding to $[(C, p)] \in \mathcal{M}_{g,1}$ is the preimage of ∞). Then $\mathbb{G}_{g,d} = d^{2g}\mathbb{G}_{g,1}$. (We omit the proof, but the main idea behind this is the Fourier–Mukai fact [Lo, Lemma 2.10].)

Define the Faber–Hurwitz class $\mathbb{F}^{g,\alpha}$ as the image in $A_{2g-1}(\mathcal{M}_{g,n}^{rt})$ of

$$\cap_{i=1}^{r-(2g-1)} br^{-1}(p_i) \cap [\overline{\mathcal{M}}_{g,\alpha}(\mathbb{P}^1, d)]^{\mathrm{virt}}$$

where the p_i are general points of $\mathbb{P}^1$. (This is the image of the Hurwitz class $\mathbb{H}_{2g-1}^{g,\alpha}$ in $\mathcal{M}_{g,n}^{rt}$.)

As with the proof of Theorem $\star$, we get at this class inductively using degeneration, and connect it to intersections of ψ-classes using localization.

7.6. Degeneration

Break the target into two pieces $\mathbb{P}^1 \rightsquigarrow \mathbb{P}^1 \cup \mathbb{P}^1$, where ∞ and one p_i are on the "right" piece, and the remaining p_i's are on the "left" piece. The Faber–Hurwitz class breaks into various pieces; we enumerate the possibilities. We are interested only in components where there is a nonsingular genus g curve on one side. We have two cases, depending on whether this curve maps to the "left" or the "right" $\mathbb{P}^1$.

7.7. If it maps to the left component, then all $2g - 1$ "moving" branch points must also map to the left component in order to get a non-zero contribution in $A_*(\mathcal{M}_{g,n}^{rt})$, by Remark 7.4. Thus by Remark 7.3, the cover on the right is of a particular sort, and the cover on the left is another Faber–Hurwitz class, where one of the branch points over ∞ has been replaced two, or where two of the branch points are replaced by one.

7.8. If the genus g curve maps to the right component, then all $2g-1$ "moving" branch points must map to the right component, and by Remark 7.4 our contribution is a certain multiple of $\mathbb{G}_{g,d}$, which by Remark 7.5 is a certain multiple of $\mathbb{G}_{g,1}$. The contribution from the left is the genus 0 Hurwitz number H_α^0, for which Hurwitz gives us an attractive formula (7).

Unwinding this gives the recursion

$$
\mathbb{F}^{g,\alpha} = \sum_{\substack{i+j=\alpha_k \\ d'+d''=d}} ij H^0_{\alpha'} \mathbb{F}^{g,\alpha''} \binom{d+l(\alpha)-2}{d'+l(\alpha')-2, d''+l(\alpha'')-1} +
$$

(15)

$$
+ \sum_{i,j} (\alpha_i + \alpha_j) \mathbb{F}^{g,\alpha'} + \sum_{i=1}^{l(\alpha)} \alpha_i^{2g+1} H^0_\alpha \mathbb{G}_{g,1}.
$$

In this formula, the contributions from paragraph Sect. 7.7 are in the first two terms on the right side of the equation, and the contributions from Sect. 7.8 are in the last. The second term on the right corresponds to where two parts α_i and α_j of α are "joined" by the nontrivial cover of the right $\mathbb{P}^1$ to yield a new partition where α_i and α_j are replaced by $\alpha_i + \alpha_j$. The first term on the right corresponds to where one part α_k of α is "cut" into two pieces i and j, forcing the curve covering the left $\mathbb{P}^1$ to break into two pieces, one of genus 0 (corresponding to partition α') and one of genus g (corresponding to α''). The binomial coefficient corresponds to the fact that the $d + l(\alpha) - 2$ fixed branch points p_1, p_2, ...on the left component must be split between these two covers.

The base case is $\mathbb{F}^{g,(1)} = \mathbb{G}_{g,1}$. Hence we have shown that $\mathbb{F}^{g,\alpha}$ is always a multiple of $\mathbb{G}_{g,1}$, and the theory of cut-and-join type equations (developed notably by Goulden and Jackson) can be applied to solve for $\mathbb{F}^{g,\alpha}$ (in generating function form) quite explicitly.

7.9. Localization

We now get at the Faber–Hurwitz class by localizing. As with the proof of Theorem $\star$, we choose a linearization on $br^{-1}(p_i)$ that corresponds to requiring all the p_i to move to 0. We now describe the fixed loci that contribute. We won't worry about the precise contribution of each fixed locus; the important thing is to see the shape of the formula.

First note that as we have only $2g - 1$ moving branch points, in any fixed locus in the "rational-tails" locus, our genus g component cannot map to ∞, and thus must be contracted to 0. The fixed locus can certainly have genus 0 components mapping to sprouted T_i over ∞, as well as genus 0 components contracted to 0.

We now look at the contribution of this fixed locus, via the relative virtual localization formula. We will get a sum of classes glued together from various moduli spaces appearing in the description of the fixed locus (cf. Sect. 5.6). Say the contracted genus g curve meets m trivial covers, of degree β_1, ..., β_m respectively. Then the contribution from this component will be some summand of

$$
\frac{1 - \lambda_1 + \cdots + (-1)^g \lambda_g}{(1 - \beta_1 \psi_1) \cdots (1 - \beta_m \psi_m)}
$$

where ψ_i are the ψ-classes on $\overline{\mathcal{M}}_{g,m}$. Thus the contribution from this component is visibly tautological, and by Remark 7.1 the contribution will be zero if the dimension of the class is less than $2g - 1$. As the total contribution of this fixed locus is $2g - 1$, any non-zero contribution must correspond to a dimension $2g - 1$ tautological class on $\overline{\mathcal{M}}_{g,m}$ glued to a dimension 0 class on the other moduli spaces appearing in this fixed locus. This can be readily computed; the genus 0 components contracted to 0 yield binomial coefficients, any components over ∞_X turn out to yield products of *genus 0 double Hurwitz numbers*, which count branched covers of $\mathbb{P}^1$ by a genus 0 curve, with specified branching α and β above two points, and the remaining branching fixed and simple.

Equipped with this localization formula, even without worrying about the specific combinatorics, we may show the following.

7.10. Theorem. *For any n, and $\beta \vdash d$, $\pi_* \psi_1^{\beta_1} \cdots \psi_n^{\beta_n}$ is a multiple of $\mathbb{G}_{d,1}$, where π is the forgetful map to $\mathcal{M}_{g,1}$.*

We have thus fully shown the "vanishing" (or socle) part of Faber's conjecture for curves with rational tails. (This may certainly be shown by other means.) In particular, we have completed a proof of Looijenga's Theorem 4.5.

Proof. Call such a class an n-point class. We will show that such a class is a multiple of
$G_{d,1}$ modulo m-point classes, where $m < n$; the result then follows by induction. As with the proof of Theorem $\star$, we consider $\mathbb{F}^{g,\alpha}$ as α varies over all partitions of length n. Each such Faber–Hurwitz class is a multiple of $\mathbb{G}_{d,1}$ by our degeneration analysis. By our localization analysis, all of the fixed points for $\mathbb{F}^{g,\alpha}$ yield m-point classes where $m < n$ except for one, corresponding to the picture in Fig. 12. The contribution of this component is some known multiple of a polynomial in $\alpha_1, \ldots, \alpha_n$. The highest-degree coefficients of this polynomial are the n-point classes, the monomials in ψ-classes that we seek. By taking a suitable linear combination of values of the polynomial (i.e. Faber–Hurwitz classes, modulo m-point classes where $m < n$), we can obtain any coefficient, and in particular, the leading coefficients. □

A related observation is that we have now given an explicit combinatorial description of the monomials in ψ-classes, as a multiple of our generator $\mathbb{G}_{g,1}$. (In truth, we have not been careful in this exposition in describing all the combinatorial factors. See [GJV3] for a precise description.)

This combinatorialization can be made precise as follows. We create a generating function $\mathbb{F}$ for Faber–Hurwitz classes. The join-cut equation (15) allows us to solve for the generating function $\mathbb{F}$.

We make a second generating function $\mathbb{W}$ for the intersections $\pi_* \psi_1^{\beta_1} \cdots \psi_n^{\beta_n} \lambda_k \in R_{2g-1}(\mathcal{M}_{g,1})$ (where $\beta_1 + \cdots + \beta_n + k = g - 2$). Localization gives us a description of $\mathbb{F}$ in terms of $\mathbb{W}$ (and also the genus 0 double Hurwitz generating function). By inverting this relationship we can hope to solve relatively explicitly for $\mathbb{W}$, and hence prove Faber's intersection number conjecture. Because genus 0 double Hurwitz number $H_{\alpha,\beta}^0$ are only currently well-understood when one of the partition has at most 3 parts (see [GJV2]), this program is not yet complete. However, it indeed yields:

7.11. *Theorem [GJV3]. Faber's intersection number conjecture is true for up to three points.*

One might reasonably hope that this will give an elegant proof of Faber's intersection number conjecture in full before long.

8 Conclusion

In the last fifteen years, there has been a surge of progress in understanding curves and their moduli using the techniques of Gromov–Witten theory. Many of these techniques have been outlined here.

Although this recent progress uses very modern machinery, it is part of an ancient story. Since the time of Riemann, algebraic curves have been studied by way of branched covers of $\mathbb{P}^1$. The techniques described here involve thinking about curves in the same way. Gromov–Witten theory gives the added insight that we should work with a "compactification" of the space of branched covers, the moduli space of stable (relative) maps. A priori we pay a steep price, by working with a moduli space that is bad in all possible ways (singular, reducible, not even equidimensional). But it is in some sense "virtually smooth", and its virtual fundamental class behaves very well, in particular with respect to degeneration and localization.

The approaches outlined here have one thing in common: in each case the key idea is direct and naive. Then one works to develop the necessary Gromov–Witten-theoretic tools to make the naive idea precise.

In conclusion, the story of using Gromov–Witten theory to understand curves, and to understand curves by examining how they map into other spaces (such as $\mathbb{P}^1$), is most certainly not over, and may just be beginning.

References

[ArbC]   E. Arbarello and M. Cornalba, *The Picard groups of the moduli spaces of curves*, Topology **26** (1987), 153–171.

[ArcL1]   D. Arcara and Y.-P. Lee, *Tautological equations in genus 2 via invariance conjectures*, preprint 2005, math.AG/0502488.

[ArcL2]   D. Arcara and Y.-P. Lee, *Tautological equations in $\overline{M}_{3,1}$ via invariance conjectures*, preprint 2005, math.AG/0503184.

[ArcS]   D. Arcara and F. Sato, *Recursive formula for $\psi^g - la_1\psi^{g-1} + \cdots + (-1)^g\lambda_g$ in $\overline{M}_{g,1}$*, preprint 2006, math.AG/0605343.

[Arn]   V. I. Arnol'd, *Topological classification of trigonometric polynomials and combinatorics of graphs with an equal number of vertices and edges*, Funct. Anal. and its Appl. **30** no. 1 (1996), 1–17.

[AB]   M. Atiyah and R. Bott, *The moment map and equivariant cohomology*, Topology **23** (1984), 1–28.

[BP]   P. Belorousski and R. Pandharipande, *A descendent relation in genus 2*, Ann. Scuola Norm. Sup. Pisa Cl. Sci. (4) **29** (2000), no. 1, 171–191.

[C]   K. Costello, *Higher-genus Gromov–Witten invariants as genus 0 invariants of symmetric products*, preprint 2003, math.AG/0303387.

[CK]   D. Cox and S. Katz, *Mirror Symmetry and Algebraic Geometry*, Mathematical surveys and Monographs **68**, Amer. Math. Soc., Providence, RI, 1999.

[CT]   M. Crescimanno and W. Taylor, *Large N phases of chiral QCD_2*, Nuclear Phys. B **437** (1995), 3–24.

[D]   J. Dénes, *The representation of a permutation as the product of a minimal number of transpositions and its connection with the theory of graphs*, Publ. Math. Ins. Hungar. Acad. Sci. **4** (1959), 63–70.

[DVV]   R. Dijkgraaf, H. Verlinde, and E. Verlinde, *Topological strings in $d < 1$*, Nuclear Phys. B **352** (1991), 59–86.

[E]   D. Edidin, *Notes on the construction of the moduli space of curves*, in *Recent progress in intersection theory (Bologna, 1997)*, 85–113, Trends Math., Birkhäuser Boston, Boston, MA, 2000.

[ELSV1]   T. Ekedahl, S. Lando, M. Shapiro, and A. Vainshtein, *On Hurwitz numbers and Hodge integrals*, C. R. Acad. Sci. Paris Sér. I Math. **328** (1999), 1175–1180.

[ELSV2]   T. Ekedahl, S. Lando, M. Shapiro, and A. Vainshtein, *Hurwitz numbers and intersections on moduli spaces of curves*, Invent. Math. **146** (2001), 297–327.

[Fab1]   C. Faber, *A conjectural description of the tautological ring of the moduli space of curves*, in *Moduli of Curves and Abelian Varieties*, 109–129, Aspects Math., **E33**, Vieweg, Braunschweig, 1999.

[Fab2]   C. Faber, MAPLE program for computing Hodge integrals, personal communication. Available at `http://math.stanford.edu/~vakil/programs/`.

[Fab3]   C. Faber, *Algorithms for computing intersection numbers on moduli spaces of curves, with an application to the class of the locus of Jacobians*, in *New trends in algebraic geometry (Warwick, 1996)*, 93–109, London Math. Soc. Lecture Note Ser., 264, Cambridge Univ. Press, Cambridge, 1999.

[Fab4]   C. Faber, personal communication, January 8, 2006.

[FabP1]   C. Faber and R. Pandharipande, *Logarithmic series and Hodge integrals in the tautological ring*, Michigan Math. J. (Fulton volume) **48** (2000), 215–252.

[FabP2]   C. Faber and R. Pandharipande, *Hodge integrals, partition matrices, and the λ_g conjecture*, Ann. Math. **157** (2003), 97–124.

[FabP3]   C. Faber and R. Pandharipande, *Relative maps and tautological classes*, J. Eur. Math. Soc. **7** (2005), no. 1, 13–49.

[Fan]   B. Fantechi, *Stacks for everybody*, in *European Congress of Mathematics, Vol. I (Barcelona, 2000)*, 349–359, Progr. Math., **201**, Birkhäuser, Basel, 2001.

[FanP]   B. Fantechi and R. Pandharipande, *Stable maps and branch divisors*, Compositio Math. **130** (2002), no. 3, 345–364.

[Ga]   A. Gathmann, *Absolute and relative Gromov–Witten invariants of very ample hypersurfaces*, Duke Math. J. **115** (2002), 171–203.

[Ge1]   E. Getzler, *Intersection theory on $\overline{M}_{1,4}$ and elliptic Gromov–Witten invariants*, J. Amer. Math. Soc. **10** (1997), no. 4, 973–998.

[Ge2]   E. Getzler, *Topological recursion relations in genus 2*, in *Integrable systems and algebraic geometry (Kobe/Kyoto, 1997)*, 73–106, World Sci. Publishing, River Edge, NJ, 1998.

[GeP]   E. Getzler and R. Pandharipande, *Virasoro constraints and the Chern classes of the Hodge bundle*, Nuclear Phys. B **530** (1998), 701–714.

[GiaM]   J. Giansiracusa and D. Maulik, *Topology and geometry of the moduli space of curves*, on-line collection of resources on moduli of curves and related subjects, http://www.aimath.org/WWN/modspacecurves/.

[Giv]   A. Givental, *Gromov–Witten invariants and quantization of quadratic hamiltonians*, Mosc. Math. J. **1** (2001), no. 4, 551–568, 645.

[GJ1]   I. P. Goulden and D. M. Jackson, *Transitive factorizations into transpositions and holomorphic mappings on the sphere*, Proc. Amer. Math. Soc. **125** (1997), 51–60.

[GJ2]   I. P. Goulden and D. M. Jackson, *The number of ramified coverings of the sphere by the double torus, and a general form for higher genera*, J. Combin. Theory A **88** (1999) 259–275.

[GJV1]   I. P. Goulden, D. M. Jackson, R. Vakil, *The Gromov–Witten potential of a point, Hurwitz numbers, and Hodge integrals*, Proc. London Math. Soc. (3) **83** (2001), 563–581.

[GJV2]   I. P. Goulden, D. M. Jackson, R. Vakil, *Towards the geometry of double Hurwitz numbers*, Adv. Math. (Artin issue) **198** (2005), 43–92.

[GJV3]   I. P. Goulden, D. M. Jackson, R. Vakil, *On Faber's intersection number conjecture on the moduli space of curves*, in preparation.

[GrP]   T. Graber and R. Pandharipande, *Localization of virtual classes*, Invent. Math. **135** (1999), 487–518.

[GrV1]   T. Graber and R. Vakil, *Hodge integrals and Hurwitz numbers via virtual localization*, Compositio Math. **135** (2003), no. 1, 25–36.

[GrV2]   T. Graber and R. Vakil, *On the tautological ring of $\overline{M}_{g,n}$*, in *Proceedings of the Seventh Gökova Geometry-Topology Conference 2000*, International Press, 2000.

[GrV3]   T. Graber and R. Vakil, *Relative virtual localization and vanishing of tautological classes on moduli spaces of curves*, Duke Math. J. **130** (2005), no. 1, 1–37.

[HLo]   R. Hain and E. Looijenga, *Mapping class groups and moduli spaces of curves*, Proc. Sympos. Pure Math. **62** Part 2, pp. 97–142, Amer. Math. Soc., Providence, RI, 1997.

[HZ]   J. Harer and D. Zagier, *The Euler characteristic of the moduli space of curves*, Invent. Math. **85** (1986), no. 3, 457–485.

[HM]   J. Harris and I. Morrison, *Moduli of Curves*, Graduate Texts in Mathematics **187**, Springer-Verlag, New York, 1998.

[HKKPTVVZ]   K. Hori, S. Katz, A. Klemm, R. Pandharipande, R. Thomas, C. Vafa, R. Vakil, and E. Zaslow, *Gromov–Witten Theory and Mirror Symmetry*, Clay Math. Inst., Amer. Math. Soc., 2002.

[H]   A. Hurwitz, *Über Riemann'sche Flächen mit gegeben Verzwei-gungspunkten*, Math. Ann. **39** (1891), 1–60.

[I1]   E. Ionel, *Topological recursive relations in $H^{2g}(M_{g,n})$*, Invent. Math. **148** (2002), no. 3, 627–658.

[I2]   E. Ionel, *Relations in the tautological ring of M_g*, Duke Math. J. **129** (2005), no. 1, 157–186.

[IP1]   E.-N. Ionel and T. H. Parker, *Relative Gromov–Witten invariants*, Ann. of Math. (2) **157** (2003), 45–96.

[IP2]   E.-N. Ionel and T. H. Parker, *The symplectic sum formula for Gromov–Witten invariants*, Ann. of Math. (2), **159** (2004), 935–1025.

[Ke]   S. Keel, *Intersection theory of moduli space of stable n-pointed curves of genus zero*, Trans. Amer. Math. Soc. **330** (1992), no. 2, 545–574.

[KiL]   Y.-S. Kim and K. Liu, *A simple proof of Witten conjecture through localization*, preprint 2005, math.AG/0508384.

[KL]   T. Kimura and X. Liu, *A genus-3 topological recursion relation*, preprint 2005, math.DG/0502457.

[Ko1]   M. Kontsevich, *Intersection theory on the moduli space of curves and the matrix Airy function*, Comm. Math. Phys. **147** (1992), 1–23.

[Ko2]   M. Kontsevich, *Enumeration of rational curves via torus actions*, in *the Moduli Space of Curves (Texel Island, 1994)*, R. Dijkgraaf, C. Faber and G. van der Geer, eds., Progr. Math. vol. 129, Birkhäuser, Boston, 1995, pp. 335–368.

[Kr]   A. Kresch, *Cycle groups for Artin stacks*, Invent. Math. **138** (1999), no. 3, 495–536.

[Lee1]   Y.-P. Lee, *Invariance of tautological equations I: Conjectures and applications*, preprint 2006, math.AG/0604318v2.

[Lee2]   Y.-P. Lee, *Invariance of tautological equations II: Gromov–Witten theory*, preprint 2006, available at http://www.math.utah.edu/~yplee/research/.

[LeeP]   Y.-P. Lee and R. Pandharipande, *Frobenius manifolds, Gromov–Witten theory, and Virasoro constraints*, book in preparation.

[Li1]   J. Li, *Stable morphisms to singular schemes and relative stable morphisms*, J. Diff. Geom. **57** (2001), 509–578.

[Li2]   J. Li, *A degeneration formula of GW-invariants*, J. Diff. Geom. **60** (2002), 199–293.

[LR]   A.-M. Li and Y. Ruan, *Symplectic surgery and Gromov–Witten invariants of Calabi-Yau 3-folds*, Invent. Math. **145** (2001), 151–218.

[LLZ]   C.-C. M. Liu, K. Liu, and J. Zhou, *A proof of a conjecture of Marino-Vafa on Hodge integrals*, J. Diff. Geom. **65** (2004), 289–340.

[Lo]   E. Looijenga, *On the tautological ring of $\mathcal{M}_g$*, Invent. Math. **121** (1995), no. 2, 411–419.

[MT]   I. Madsen and U. Tillmann, *The stable mapping class group and $Q(\mathbb{CP}^{\infty}_+)$*, Invent. Math. **145** (2001), no. 3, 509–544.

[MW]   I. Madsen and M. Weiss, *The stable moduli space of Riemann surfaces: Mumford's conjecture*, preprint 2002, math.AT/0212321.

[Mi]   M. Mirzakhani, *Weil-Petersson volumes and the Witten-Kontsevich formula*, preprint 2003.

[Mo1]   S. Morita, *Generators for the tautological algebra of the moduli space of curves*, Topology **42** (2003), 787–819.

[Mo2]   S. Morita, *Cohomological structure of the mapping class group and beyond*, preprint 2005, math.GT/0507308v1.

[Mu]   D. Mumford, *Toward an enumerative geometry of the moduli space of curves*, in *Arithmetic and Geometry*, Vol. II, M. Artin and J. Tate ed., 271–328, Prog. Math. **36**, Birk. Boston, Boston, MA, 1983.

[OP]   A. Okounkov and R. Pandharipande, *Gromov–Witten theory, Hurwitz numbers, and matrix models, I*, math.AG/0101147.

[P]   R. Pandharipande, *Three questions in Gromov–Witten theory*, in *Proceedings of the International Congress of Mathematicians, Vol. II (Beijing, 2002)*, 503–512, Higher Ed. Press, Beijing, 2002.

[R]   B. Riemann, *Theorie der Abel'schen Funktionen*, J. Reine angew. Math. **54** (1857), 115–155.

[T]   U. Tillmann, *Strings and the stable cohomology of mapping class groups*, in *Proceedings of the International Congress of Mathematicians, Vol. II (Beijing, 2002)*, 447–456, Higher Ed. Press, Beijing, 2002.

[V1]   R. Vakil, *Genus 0 and 1 Hurwitz numbers: Recursions, formulas, and graph-theoretic interpretations*, Trans. Amer. Math. Soc. **353** (2001), 4025–4038.

[V2]   R. Vakil, *The moduli space of curves and its tautological ring*, Notices of the Amer. Math. Soc. (feature article), vol. 50, no. 6, June/July 2003, p. 647–658.

[Vi]   A. Vistoli, *Intersection theory on algebraic stacks and on their moduli spaces*, Invent. Math. **97** (1989), 613–670.

[W]   E. Witten, *Two dimensional gravity and intersection theory on moduli space*, Surveys in Diff. Geom. **1** (1991), 243–310.

List of Participants

1. Abramovich Dan
   Brown Univ., Providence,
   USA
   abrmovich@math.bu.edu
   (**lecturer**)
2. Abriani Devis
   SISSA, Trieste, Italy
   abriani@sissa.it
3. Alexandrov Alexander
   MIPT, Moscow, Russia
   al@itep.ru
4. Amburg Natalia
   ITEP, Moscow, Russia
   amburg@itep.ru
5. Andreini Elena
   SISSA, Trieste, Italy
   andreini@sissa.it
6. Bartocci Claudio
   Univ. of Genova, Genova,
   Italy
   bartocci@dima.unige.it
7. Behrend Kai
   Univ. of British Columbia
   Vancouver, Canada
   behrend@math.ubc.ca
   (**editor**)
8. Bergstrom Jonas
   Royal Inst. Tech
   Stockholm, Sweden
   jonasb@math.kth.se
9. Bisi Cinzia
   Univ. of Calabria, Cosenza,
   Italy
   bisi@math.unifi.it
10. Blasi Francesco Simone
    Univ. of Roma Tor Vergata Rome,
    Italy
    blasi@mat.uniroma2.it
11. Bruno Andrea
    Univ. of Roma3, Rome,
    Italy
    bruno@mat.uniroma3.it
12. Bryan Jim
    Univ. of British Columbia
    Vancouver, Canada
    jbryan@math.ubc.ca
13. Cavalieri Renzo
    Univ. of Utah
    Salt Lake City, USA
    renzo@math.utah.edu
14. D'Agnolo Andrea
    Univ. of Padova, Padua,
    Italy
    dagnolo@math.unipd.it
15. Dionisi Carla
    Univ. of Firenze, Florence,
    Italy
    dionisi@math.unifi.it

16. Donati Fabrizio
SISSA, Trieste, Italy
donati@sissa.it

17. Fantechi Barbara
SISSA, Trieste, Italy
fantechi@sissa.it

18. Fiorenza Domenico
Univ. of Roma "La Sapienza"
Rome, Italy
fiorenza@mat.uniroma1.it

19. Fulghesu Damiano
Univ. of Bologna, Bologna,
Italy
fulghesu@dm.unibo.it

20. Gholampour Amin
Univ. of British Columbia
Vancouver, Canada
amin@math.ubc.ca

21. Gonzalez Eduardo
SUNY, Stony Brook, USA
eduardo@math.sunysb.edu

22. Gorskiy Evgeny
Independent Univ.
Moscow, Russia
gorsky@mccme.ru

23. Gromov Nikolay
St Petersburg Univ.
St Petersburg, Russia
nik_gromov@mail.ru

24. Kerr Gabriel
Univ. of Chicago, Chicago,
USA
gdkerr@math.uchicago.edu

25. Kobak Dmitry
St Petersburg Univ.
St Petersburg, Russia
dkobak@gmail.com

26. Manetti Marco
Univ. of Roma "La Sapienza"
Rome, Italy
manetti@mat.uniroma1.it
(**editor**)

27. Mann Etienne
St Petersburg Univ.
St Petersburg, Russia
mann@math.u-strasbg.fr

28. Mariño Marcos
CERN, Geneva, Switzerland
marcosm@physics.rutgers.edu
(**lecturer**)

29. Matsumura Tomoo
Boston Univ., Boston, USA
mushmt@math.bu.edu

30. Mencattini Igor
St Petersburg Univ.
St Petersburg, Russia
igorre@math.bu.edu

31. Migliorini Luca
Univ. of Bologna, Bologna,
Italy
migliori@dm.unibo.it

32. Mistretta Carlo Ernesto
Univ. of Paris7, Paris, France
ernesto@math.jussieu.fr

33. Mnev Pavel
St Petersburg Univ.
St Petersburg, Russia
pasha_mnev@mail.ru

34. Noseda Francesco
SISSA, Trieste, Italy
noseda@sissa.it

35. Ottaviani Giorgio
Univ. of Firenze, Florence,
Italy
ottavian@math.unifi.it

36. Perroni Fabio
SISSA, Trieste, Italy
perroni@sissa.it

37. Przyjalkovski Victor
Steklov Math. Inst.
Moscow, Russia
victorprz@mi.ras.ru

38. Ricco Antonio
SISSA, Trieste, Italy
ricco@sissa.it

39. Ruddat Helge
    Freiburg Univ.
    Freiburg, Germany
    helge.ruddat@gmx.de

40. Rybnikov Leonid
    Moscow State Univ.
    Moscow, Russia
    leo_rybnikov@mtu-net.ru

41. Justin Sawon
    SUNY, Stony Brook,
    USA
    sawon@math.sunysb.edu

42. Sernesi Edoardo
    Univ. of Roma3, Rome,
    Italy
    sernesi@matrm3.mat.uniroma3.it

43. Stellari Paolo
    Univ. of Milano, Milan, Italy
    stellari@mat.unimi.it

44. Szendroi Balazs
    Univ. of Utrecht
    Utrecht, The Netherlands
    szendroi@math.uu.nl

45. Thaddeus Michael
    Columbia Univ., New York, USA
    thaddeus@math.columbia.edu
    (**lecturer**)

46. Thier Christian
    Freiburg Univ.
    Freiburg, Germany
    chthier@gmx.de

47. Vakil Ravi
    Stanford Univ., Stanford, USA
    vakil@math.stanford.edu
    (**lecturer**)

48. Vassiliev Dmitry
    ITEP, Moscow, Russia
    vasiliev@itap.ru

49. Vetro Francesca
    Univ. of Palermo, Palermo,
    Italy
    fvetro@math.unipa.it

50. Watanabe Satoru
    Univ. of Paris VI, Paris,
    France
    watanabe@math.jussieu.fr

LIST OF C.I.M.E. SEMINARS

Published by C.I.M.E

Published by Ed. Cremonese, Firenze

1966   39. Calculus of variations
40. Economia matematica
41. Classi caratteristiche e questioni connesse
42. Some aspects of diffusion theory

1967   43. Modern questions of celestial mechanics
44. Numerical analysis of partial differential equations
45. Geometry of homogeneous bounded domains

1968   46. Controllability and observability
47. Pseudo-differential operators
48. Aspects of mathematical logic

1969   49. Potential theory
50. Non-linear continuum theories in mechanics and physics and their applications
51. Questions of algebraic varieties

1970   52. Relativistic fluid dynamics
53. Theory of group representations and Fourier analysis
54. Functional equations and inequalities
55. Problems in non-linear analysis

1971   56. Stereodynamics
57. Constructive aspects of functional analysis (2 vol.)
58. Categories and commutative algebra

1972   59. Non-linear mechanics
60. Finite geometric structures and their applications
61. Geometric measure theory and minimal surfaces

1973   62. Complex analysis
63. New variational techniques in mathematical physics
64. Spectral analysis

1974   65. Stability problems
66. Singularities of analytic spaces
67. Eigenvalues of non linear problems

1975   68. Theoretical computer sciences
69. Model theory and applications
70. Differential operators and manifolds

Published by Ed. Liguori, Napoli

1976   71. Statistical Mechanics
72. Hyperbolicity
73. Differential topology

1977   74. Materials with memory
75. Pseudodifferential operators with applications
76. Algebraic surfaces

Published by Ed. Liguori, Napoli & Birkhäuser

1978   77. Stochastic differential equations
78. Dynamical systems

1979   79. Recursion theory and computational complexity
80. Mathematics of biology

1980   81. Wave propagation
82. Harmonic analysis and group representations
83. Matroid theory and its applications

Published by Springer-Verlag

Lecture Notes in Mathematics

For information about earlier volumes
please contact your bookseller or Springer
LNM Online archive: springerlink.com

Vol. 1807: V. D. Milman, G. Schechtman (Eds.), Geometric Aspects of Functional Analysis. Israel Seminar 2000-2002 (2003)

Vol. 1808: W. Schindler, Measures with Symmetry Properties (2003)

Vol. 1809: O. Steinbach, Stability Estimates for Hybrid Coupled Domain Decomposition Methods (2003)

Vol. 1810: J. Wengenroth, Derived Functors in Functional Analysis (2003)

Vol. 1811: J. Stevens, Deformations of Singularities (2003)

Vol. 1812: L. Ambrosio, K. Deckelnick, G. Dziuk, M. Mimura, V. A. Solonnikov, H. M. Soner, Mathematical Aspects of Evolving Interfaces. Madeira, Funchal, Portugal 2000. Editors: P. Colli, J. F. Rodrigues (2003)

Vol. 1813: L. Ambrosio, L. A. Caffarelli, Y. Brenier, G. Buttazzo, C. Villani, Optimal Transportation and its Applications. Martina Franca, Italy 2001. Editors: L. A. Caffarelli, S. Salsa (2003)

Vol. 1814: P. Bank, F. Baudoin, H. Föllmer, L.C.G. Rogers, M. Soner, N. Touzi, Paris-Princeton Lectures on Mathematical Finance 2002 (2003)

Vol. 1815: A. M. Vershik (Ed.), Asymptotic Combinatorics with Applications to Mathematical Physics. St. Petersburg, Russia 2001 (2003)

Vol. 1816: S. Albeverio, W. Schachermayer, M. Talagrand, Lectures on Probability Theory and Statistics. Ecole d'Eté de Probabilités de Saint-Flour XXX-2000. Editor: P. Bernard (2003)

Vol. 1817: E. Koelink, W. Van Assche (Eds.), Orthogonal Polynomials and Special Functions. Leuven 2002 (2003)

Vol. 1818: M. Bildhauer, Convex Variational Problems with Linear, nearly Linear and/or Anisotropic Growth Conditions (2003)

Vol. 1819: D. Masser, Yu. V. Nesterenko, H. P. Schlickewei, W. M. Schmidt, M. Waldschmidt, Diophantine Approximation. Cetraro, Italy 2000. Editors: F. Amoroso, U. Zannier (2003)

Vol. 1820: F. Hiai, H. Kosaki, Means of Hilbert Space Operators (2003)

Vol. 1821: S. Teufel, Adiabatic Perturbation Theory in Quantum Dynamics (2003)

Vol. 1822: S.-N. Chow, R. Conti, R. Johnson, J. Mallet-Paret, R. Nussbaum, Dynamical Systems. Cetraro, Italy 2000. Editors: J. W. Macki, P. Zecca (2003)

Vol. 1823: A. M. Anile, W. Allegretto, C. Ringhofer, Mathematical Problems in Semiconductor Physics. Cetraro, Italy 1998. Editor: A. M. Anile (2003)

Vol. 1824: J. A. Navarro González, J. B. Sancho de Salas, $\mathscr{C}^\infty$ – Differentiable Spaces (2003)

Vol. 1825: J. H. Bramble, A. Cohen, W. Dahmen, Multiscale Problems and Methods in Numerical Simulations, Martina Franca, Italy 2001. Editor: C. Canuto (2003)

Vol. 1826: K. Dohmen, Improved Bonferroni Inequalities via Abstract Tubes. Inequalities and Identities of Inclusion-Exclusion Type. VIII, 113 p, 2003.

Vol. 1827: K. M. Pilgrim, Combinations of Complex Dynamical Systems. IX, 118 p, 2003.

Vol. 1828: D. J. Green, Gröbner Bases and the Computation of Group Cohomology. XII, 138 p, 2003.

Vol. 1829: E. Altman, B. Gaujal, A. Hordijk, Discrete-Event Control of Stochastic Networks: Multimodularity and Regularity. XIV, 313 p, 2003.

Vol. 1830: M. I. Gil', Operator Functions and Localization of Spectra. XIV, 256 p, 2003.

Vol. 1831: A. Connes, J. Cuntz, E. Guentner, N. Higson, J. E. Kaminker, Noncommutative Geometry, Martina Franca, Italy 2002. Editors: S. Doplicher, L. Longo (2004)

Vol. 1832: J. Azéma, M. Émery, M. Ledoux, M. Yor (Eds.), Séminaire de Probabilités XXXVII (2003)

Vol. 1833: D.-Q. Jiang, M. Qian, M.-P. Qian, Mathematical Theory of Nonequilibrium Steady States. On the Frontier of Probability and Dynamical Systems. IX, 280 p, 2004.

Vol. 1834: Yo. Yomdin, G. Comte, Tame Geometry with Application in Smooth Analysis. VIII, 186 p, 2004.

Vol. 1835: O.T. Izhboldin, B. Kahn, N.A. Karpenko, A. Vishik, Geometric Methods in the Algebraic Theory of Quadratic Forms. Summer School, Lens, 2000. Editor: J.-P. Tignol (2004)

Vol. 1836: C. Năstăsescu, F. Van Oystaeyen, Methods of Graded Rings. XIII, 304 p, 2004.

Vol. 1837: S. Tavaré, O. Zeitouni, Lectures on Probability Theory and Statistics. Ecole d'Eté de Probabilités de Saint-Flour XXXI-2001. Editor: J. Picard (2004)

Vol. 1838: A.J. Ganesh, N.W. O'Connell, D.J. Wischik, Big Queues. XII, 254 p, 2004.

Vol. 1839: R. Gohm, Noncommutative Stationary Processes. VIII, 170 p, 2004.

Vol. 1840: B. Tsirelson, W. Werner, Lectures on Probability Theory and Statistics. Ecole d'Eté de Probabilités de Saint-Flour XXXII-2002. Editor: J. Picard (2004)

Vol. 1841: W. Reichel, Uniqueness Theorems for Variational Problems by the Method of Transformation Groups (2004)

Vol. 1842: T. Johnsen, A. L. Knutsen, K_3 Projective Models in Scrolls (2004)

Vol. 1843: B. Jefferies, Spectral Properties of Noncommuting Operators (2004)

Vol. 1844: K.F. Siburg, The Principle of Least Action in Geometry and Dynamics (2004)

Vol. 1845: Min Ho Lee, Mixed Automorphic Forms, Torus Bundles, and Jacobi Forms (2004)

Vol. 1846: H. Ammari, H. Kang, Reconstruction of Small Inhomogeneities from Boundary Measurements (2004)

Vol. 1847: T.R. Bielecki, T. Björk, M. Jeanblanc, M. Rutkowski, J.A. Scheinkman, W. Xiong, Paris-Princeton Lectures on Mathematical Finance 2003 (2004)

Vol. 1848: M. Abate, J. E. Fornaess, X. Huang, J. P. Rosay, A. Tumanov, Real Methods in Complex and CR Geometry, Martina Franca, Italy 2002. Editors: D. Zaitsev, G. Zampieri (2004)

Vol. 1849: Martin L. Brown, Heegner Modules and Elliptic Curves (2004)

Vol. 1850: V. D. Milman, G. Schechtman (Eds.), Geometric Aspects of Functional Analysis. Israel Seminar 2002-2003 (2004)

Vol. 1851: O. Catoni, Statistical Learning Theory and Stochastic Optimization (2004)

Vol. 1852: A.S. Kechris, B.D. Miller, Topics in Orbit Equivalence (2004)

Vol. 1853: Ch. Favre, M. Jonsson, The Valuative Tree (2004)

Vol. 1854: O. Saeki, Topology of Singular Fibers of Differential Maps (2004)

Vol. 1855: G. Da Prato, P.C. Kunstmann, I. Lasiecka, A. Lunardi, R. Schnaubelt, L. Weis, Functional Analytic Methods for Evolution Equations. Editors: M. Iannelli, R. Nagel, S. Piazzera (2004)

Vol. 1856: K. Back, T.R. Bielecki, C. Hipp, S. Peng, W. Schachermayer, Stochastic Methods in Finance, Bressanone/Brixen, Italy, 2003. Editors: M. Fritelli, W. Rung-galdier (2004)

Vol. 1857: M. Émery, M. Ledoux, M. Yor (Eds.), Séminaire de Probabilités XXXVIII (2005)

Vol. 1858: A.S. Cherny, H.-J. Engelbert, Singular Stochastic Differential Equations (2005)

Vol. 1859: E. Letellier, Fourier Transforms of Invariant Functions on Finite Reductive Lie Algebras (2005)

Vol. 1860: A. Borisyuk, G.B. Ermentrout, A. Friedman, D. Terman, Tutorials in Mathematical Biosciences I. Mathematical Neurosciences (2005)

Vol. 1861: G. Benettin, J. Henrard, S. Kuksin, Hamiltonian Dynamics – Theory and Applications, Cetraro, Italy, 1999. Editor: A. Giorgilli (2005)

Vol. 1862: B. Helffer, F. Nier, Hypoelliptic Estimates and Spectral Theory for Fokker-Planck Operators and Witten Laplacians (2005)

Vol. 1863: H. Führ, Abstract Harmonic Analysis of Continuous Wavelet Transforms (2005)

Vol. 1864: K. Efstathiou, Metamorphoses of Hamiltonian Systems with Symmetries (2005)

Vol. 1865: D. Applebaum, B.V. R. Bhat, J. Kustermans, J. M. Lindsay, Quantum Independent Increment Processes I. From Classical Probability to Quantum Stochastic Calculus. Editors: M. Schürmann, U. Franz (2005)

Vol. 1866: O.E. Barndorff-Nielsen, U. Franz, R. Gohm, B. Kümmerer, S. Thorbjønsen, Quantum Independent Increment Processes II. Structure of Quantum Lévy Processes, Classical Probability, and Physics. Editors: M. Schürmann, U. Franz, (2005)

Vol. 1867: J. Sneyd (Ed.), Tutorials in Mathematical Biosciences II. Mathematical Modeling of Calcium Dynamics and Signal Transduction. (2005)

Vol. 1868: J. Jorgenson, S. Lang, Pos_n(R) and Eisenstein Series. (2005)

Vol. 1869: A. Dembo, T. Funaki, Lectures on Probability Theory and Statistics. Ecole d'Eté de Probabilités de Saint-Flour XXXIII-2003. Editor: J. Picard (2005)

Vol. 1870: V.I. Gurariy, W. Lusky, Geometry of Müntz Spaces and Related Questions. (2005)

Vol. 1871: P. Constantin, G. Gallavotti, A.V. Kazhikhov, Y. Meyer, S. Ukai, Mathematical Foundation of Turbulent Viscous Flows, Martina Franca, Italy, 2003. Editors: M. Cannone, T. Miyakawa (2006)

Vol. 1872: A. Friedman (Ed.), Tutorials in Mathematical Biosciences III. Cell Cycle, Proliferation, and Cancer (2006)

Vol. 1873: R. Mansuy, M. Yor, Random Times and Enlargements of Filtrations in a Brownian Setting (2006)

Vol. 1874: M. Yor, M. Émery (Eds.), In Memoriam Paul-André Meyer - Séminaire de Probabilités XXXIX (2006)

Vol. 1875: J. Pitman, Combinatorial Stochastic Processes. Ecole d'Eté de Probabilités de Saint-Flour XXXII-2002. Editor: J. Picard (2006)

Vol. 1876: H. Herrlich, Axiom of Choice (2006)

Vol. 1877: J. Steuding, Value Distributions of L-Functions (2007)

Vol. 1878: R. Cerf, The Wulff Crystal in Ising and Percolation Models, Ecole d'Eté de Probabilités de Saint-Flour XXXIV-2004. Editor: Jean Picard (2006)

Vol. 1879: G. Slade, The Lace Expansion and its Applications, Ecole d'Eté de Probabilités de Saint-Flour XXXIV-2004. Editor: Jean Picard (2006)

Vol. 1880: S. Attal, A. Joye, C.-A. Pillet, Open Quantum Systems I, The Hamiltonian Approach (2006)

Vol. 1881: S. Attal, A. Joye, C.-A. Pillet, Open Quantum Systems II, The Markovian Approach (2006)

Vol. 1882: S. Attal, A. Joye, C.-A. Pillet, Open Quantum Systems III, Recent Developments (2006)

Vol. 1883: W. Van Assche, F. Marcellàn (Eds.), Orthogonal Polynomials and Special Functions, Computation and Application (2006)

Vol. 1884: N. Hayashi, E.I. Kaikina, P.I. Naumkin, I.A. Shishmarev, Asymptotics for Dissipative Nonlinear Equations (2006)

Vol. 1885: A. Telcs, The Art of Random Walks (2006)

Vol. 1886: S. Takamura, Splitting Deformations of Degenerations of Complex Curves (2006)

Vol. 1887: K. Habermann, L. Habermann, Introduction to Symplectic Dirac Operators (2006)

Vol. 1888: J. van der Hoeven, Transseries and Real Differential Algebra (2006)

Vol. 1889: G. Osipenko, Dynamical Systems, Graphs, and Algorithms (2006)

Vol. 1890: M. Bunge, J. Funk, Singular Coverings of Toposes (2006)

Vol. 1891: J.B. Friedlander, D.R. Heath-Brown, H. Iwaniec, J. Kaczorowski, Analytic Number Theory, Cetraro, Italy, 2002. Editors: A. Perelli, C. Viola (2006)

Vol. 1892: A. Baddeley, I. Bárány, R. Schneider, W. Weil, Stochastic Geometry, Martina Franca, Italy, 2004. Editor: W. Weil (2007)

Vol. 1893: H. Hanßmann, Local and Semi-Local Bifurcations in Hamiltonian Dynamical Systems, Results and Examples (2007)

Vol. 1894: C.W. Groetsch, Stable Approximate Evaluation of Unbounded Operators (2007)

Vol. 1895: L. Molnár, Selected Preserver Problems on Algebraic Structures of Linear Operators and on Function Spaces (2007)

Vol. 1896: P. Massart, Concentration Inequalities and Model Selection, Ecole d'Été de Probabilités de Saint-Flour XXXIII-2003. Editor: J. Picard (2007)

Vol. 1897: R. Doney, Fluctuation Theory for Lévy Processes, Ecole d'Été de Probabilités de Saint-Flour XXXV-2005. Editor: J. Picard (2007)

Vol. 1898: H.R. Beyer, Beyond Partial Differential Equations, On linear and Quasi-Linear Abstract Hyperbolic Evolution Equations (2007)

Vol. 1899: Séminaire de Probabilités XL. Editors: C. Donati-Martin, M. Émery, A. Rouault, C. Stricker (2007)

Vol. 1900: E. Bolthausen, A. Bovier (Eds.), Spin Glasses (2007)

Vol. 1901: O. Wittenberg, Intersections de deux quadriques et pinceaux de courbes de genre 1, Intersections of Two Quadrics and Pencils of Curves of Genus 1 (2007)

Vol. 1902: A. Isaev, Lectures on the Automorphism Groups of Kobayashi-Hyperbolic Manifolds (2007)

Vol. 1903: G. Kresin, V. Maz'ya, Sharp Real-Part Theorems (2007)

Vol. 1904: P. Giesl, Construction of Global Lyapunov Functions Using Radial Basis Functions (2007)

Vol. 1905: C. Prévôt, M. Röckner, A Concise Course on Stochastic Partial Differential Equations (2007)

Vol. 1906: T. Schuster, The Method of Approximate Inverse: Theory and Applications (2007)

Vol. 1907: M. Rasmussen, Attractivity and Bifurcation for Nonautonomous Dynamical Systems (2007)

Vol. 1908: T.J. Lyons, M. Caruana, T. Lévy, Differential Equations Driven by Rough Paths, Ecole d'Été de Probabilités de Saint-Flour XXXIV-2004 (2007)

Recent Reprints and New Editions